S0-EOJ-206

LIBRARY
AUG 7 '78
JOHN DEERE
PRODUCT ENGINEERING

ASME SI-10

STEAM CHARTS

Thermodynamic Properties of Steam, h-v, in Graphical Form
for the Superheated, Vapor and Liquid Conditions
in both SI (Metric) and U.S. Customary Units

by

JAMES H. POTTER

George Meade Bond Professor of Mechanical Engineering
Stevens Institute of Technology

THE AMERICAN SOCIETY OF MECHANICAL ENGINEERS

United Engineering Center • 345 East 47th Street • New York, N.Y. 10017

Library of Congress Catalog Card No. 75-32066

Copyright © 1976 by
THE AMERICAN SOCIETY OF MECHANICAL ENGINEERS
United Engineering Center
345 East 47 Street
New York, N.Y. 10017
Printed in U.S.A.

*

Dedicated to the memory

of

RICHARD MOLLIER

*

Professor of the Theory of Machines

and twice Rector

at the

TECHNICAL UNIVERSITY OF DRESDEN

*

A brilliant innovator in the graphical

representation of thermodynamic properties

*

Contents

Preface	vii
Acknowledgements	ix
Introduction	xi
Units and Conversion Factors	
Units	xiii
Conversion Factors	xiv
SI (Metric) Chart Index	1
SI (Metric) Charts	4
U. S. Customary Chart Index	92
U. S. Customary Charts	94
Sample Problems	124

Preface

Thirty-six years have elapsed since the publication of the last set of h-v charts for steam. In the interval, new knowledge of the properties of steam and compressed liquid water have been developed and have been incorporated in the several reports of the Sixth International Conference on the Properties of Steam.

With the introduction of the SI (Metric) units, a new and distinct need has emerged for steam properties in graphical form. As there will probably be a protracted time during which both the U.S. customary and SI (Metric) units will be in use, both are included in this book.

This compilation of charts is designed to serve the need for a compact portable volume covering the properties of superheated and wet steam, as well as compressed liquid water, in both systems of units. The book should prove useful to students and to engineers in operating, field, or design assignments.

Acknowledgements

Many people contributed to the successful completion of this project, and it is hoped that none of them shall be overlooked in an expression of the gratitude of the author. For some of the early technical advice and encouragement, thanks are due Mr. C. A. Meyer of the Westinghouse Electric Company. Professor Leo C. Pigage of the University of Illinois gave much valuable advice stemming from his experience hand-plotting the h-v steam charts of 1939 during his student days at Cornell University.

The work could not have been undertaken without the technical and financial support of three great industrial organizations: the General Electric Company, Combustion Engineering, Inc., and The Babcock & Wilcox Company. The Large Steam Turbine-Generator Department of the General Electric Company supplied the computer computation of all the data points. Support from these organizations was especially gratifying in view of the fact that, with their individual computer retrieval systems, they had nothing to gain directly from the project. The entire steam-using fraternity has been well served by this unselfish gesture of support.

Special thanks are due to Mr. C. W. Elston and Mr. R. C. Spencer at Schenectady, N. Y., for aid in the planning phases. Mr. R. B. McClintock supervised the computer effort at the General Electric Company and earned the gratitude of the plotting staff. At The Babcock & Wilcox Company, generous support was provided by Mr. G. W. Kessler and Mr. S. L. Morse. Mr. R. H. Dowhan and Mr. J. H. Fernandes, at Combustion Engineering, Incorporated, supplied ready understanding and early support.

The hand-plotting of the data was shared among undergraduate students K. Clapp, S. Hoffman, E. Liu, F. A. Morgan, C. M. Nickerson, R. Ruginis, and M. W. Wagner, aided by graduate students S. A. Robinson, B. Sertmehmetoglu, H. Tokel, and A. Yorgiadis.

It is a pleasure to record the encouragement of President K. C. Rogers at the Stevens Institute of Technology and to thank the Department of Mechanical Engineering for allotting space for the plotting staff and for the storage of the master charts.

Introduction

Experimental investigation of steam properties dates at least to James Watt and the Scottish professors of his era [1]. Some of the most significant measurements of the 19th century were made for the French government by Regnault [2]. Ewing [3] claimed that the Callendar equations of 1900 [4] made possible the subsequent development of sets of steam tables.

Mollier [5] showed the way to a major improvement by his graphical representation of steam properties, especially in the h-s plane. His steam tables and Mollier diagram [6] were based on the Callendar equations.

The rapid growth of the steam turbine led to a need for h-v charts. This was met originally by Ellenwood [7], who used twelve h-v charts to cover the range of 0.1 psia, 70% quality, to 600 psia with 520°F of superheat. The principal curves included temperature, pressure, entropy, superheat, and quality. As these were plotted to cartesian coordinates, the curves tended to change abruptly over small increments of volume. The data were taken from the Marks & Davis Steam Tables [8].

The combination of higher throttle conditions, and the development of new formulations for the properties of steam and water, led to the 1936 steam tables of Keenan & Keyes [9]. The Ellenwood & Mackey h-v charts of 1939 [10] were based on these 1936 equations. The adoption of a logarithmic scale for specific volumes improved both the appearance and the clarity of the graphs. A second edition, with only minor changes, was brought out by the same authors in 1944 [11].

The present ASME steam tables [12] were based on the 1967 IFC Formulation for Industrial Use, in conformity with the 1963 skeleton tables of the Sixth International Conference on the Properties of Steam. In 1968 additional procedures were presented by McClintock and Silvestri [13].

For this volume the calculations for steam and compressed liquid, in both systems of units, were made at the Large Steam Turbine-Generator Division of the General Electric Company. The data were then hand-plotted by students at the Stevens Institute of Technology. Graphical values of the principal variables were checked against those of the 1967 ASME steam tables and the VDI SI steam tables [14].

REFERENCES

[1] "James Watt: Craftsman & Engineer," H. W. Dickinson, Cambridge Univ. Press, 1936, pg. 35.
[2] "Mémoires de L'Académie des Sciences," M. V. Regnault, vol. 21, 1847.
[3] "Thermodynamics for Engineers," J. A. Ewing, Cambridge Univ. Press, 2 ed., 1936.
[4] "On The Thermodynamical Properties of Gases & Vapors as Deduced From a Modified Form of the Joule-Thomson Equation, With Special Reference to the Properties of Steam," H.L. Callendar, *Proc. Roy. Soc.*, vol. 67, 1900, pp. 266-286.
[5] "Neue Diagramme zur Technischen Wärmelehre," R. Mollier, Zeit. VDI, 1904, pp. 271-275.
[6] "Neue Tabellen und Diagramme für Wasserdampf," R. Mollier, Verlag von Julius Springer, Berlin - 1906.
[7] "Steam Charts," F.O. Ellenwood, John Wiley & Sons; New York - 1914.
[8] "Tables & Diagrams of the Thermal Properties of Saturated & Superheated Steam," L.S. Marks & H.N. Davis, Longmans, Green, & Co., New York - 1909.
[9] "Thermodynamic Properties of Steam," J.H. Keenan & F. G. Keyes, John Wiley & Sons, Inc., New York - 1936.
[10] "Vapor Charts," F.O. Ellenwood & C.O. Mackey, John Wiley & Sons, Inc., New York - 1939.
[11] "Thermodynamic Charts," F.O. Ellenwood & C.O. Mackey, John Wiley & Sons, Inc., New York - 1944.
[12] "Thermodynamic & Transport Properties of Steam," C.A. Meyer, R. B. McClintock, G.J. Silvestri & R.C. Spencer, ASME, New York - 1967.
[13] "Formulations & Iterative Procedures for the Calculation of Properties of Steam," R.B. McClintock & G.J. Silvestri, ASME, New York - 1968.
[14] "Properties of Water & Steam in SI Units," Ernst Schmidt (VDI), Springer-Verlag, Berlin - 1969.

Units and Conversion Factors

For a more complete treatment of SI (Metric) and U.S. customary units, the reader is referred to ASME Guide SI-1.

The following units are used in the charts:

Quantity	Symbol	SI (Metric)	U.S. Customary
Moisture		%	%
Pressure	p	Pascal	psia
Specific enthalpy	h	kJ/kg	Btu/lb
Specific entropy	s	kJ/(kg·K)	Btu/(lb(R))
Specific volume	v	(m)^3/kg	(ft)^3/lb
Temperature	t	°C	°F
Temperature, Absolute	T	K	°R

Subscripts:
    Saturated liquid   f
    Saturated vapor   g

Conversion Factors

Quantity	Given	Multiply by	To Obtain
Acceleration	ft/sec^2	3.048×10^{-1}	m/s^2
Area	in^2	6.4516×10^{-4}	m^2
Energy, Work	Btu (IT)	1.0550×10^3	J
Energy-specific	Btu/lb	2.326	kJ/kg
Heat rate	Btu/(kW (hr))	2.93071×10^{-4}	kJ/(kW·s)
Mass	lb (avoir)	4.5359×10^{-1}	kg
Power	hp (550 ft-lbf/sec)	7.457×10^{-1}	kW
Pressure	in Hg a	3.38639×10^3	Pa
Pressure	psia	6.895×10^3	Pa
Pressure	std. atm.	29.92	in Hg a
Pressure	std. atm.	1.013253×10^5	Pa
Pressure	std. atm.	14.696	psia
Thermal Efficiency		$1.0 \times$ [kJ/kW·s]$^{-1}$	%
Thermal Efficiency		3412/Btu/(kW(hr))	%
Velocity	ft/min	5.08×10^{-3}	m/s
Velocity	ft/sec	3.048×10^{-1}	m/s
Volume	ft^3	2.8317×10^{-2}	m^3
Volume	liter	1.0000×10^{-3}	m^3
Volume	U.S. gallons	3.7854×10^{-3}	m^3

Item	SI (Metric)	U.S. Customary
One standard atm.	101 325.3 Pa	29.92 in Hg a or 14.696 psia
Heat Rate	kJ/(kW·s)	Btu/(kW(hr))
Thermal Efficiency	1/(heat rate)	3412/(heat rate)

SI (METRIC) CHARTS

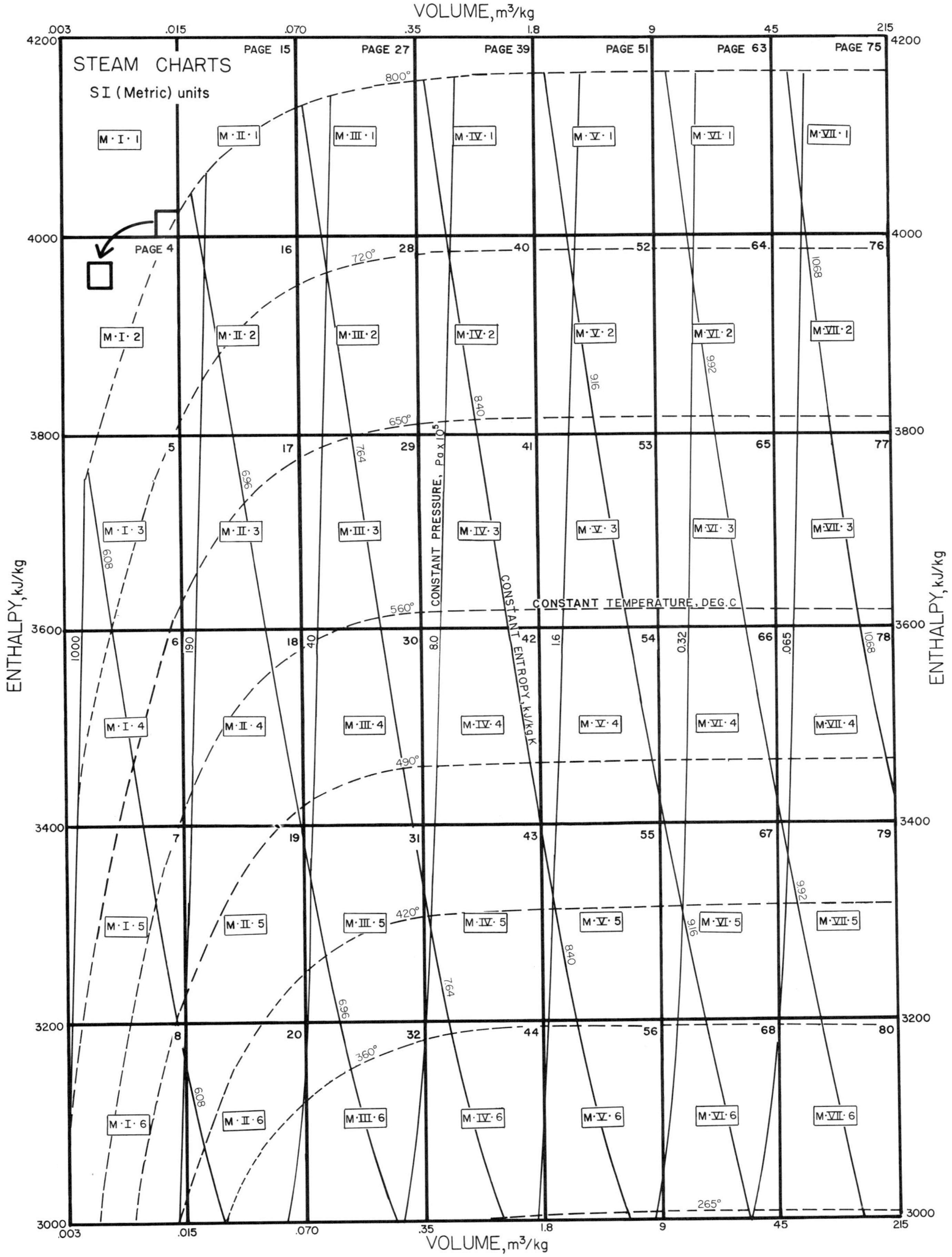
VOLUME, m³/kg
.003
.015
.070
.35
1.8
9
45
215
STEAM CHARTS
SI (Metric) units
PAGE 4
PAGE 15
PAGE 27
PAGE 39
PAGE 51
PAGE 63
PAGE 75
5
6
7
8
16
17
18
19
20
28
29
30
31
32
40
41
42
43
44
52
53
54
55
56
64
65
66
67
68
76
77
78
79
80
M·I·1
M·II·1
M·III·1
M·IV·1
M·V·1
M·VI·1
M·VII·1
M·I·2
M·II·2
M·III·2
M·IV·2
M·V·2
M·VI·2
M·VII·2
M·I·3
M·II·3
M·III·3
M·IV·3
M·V·3
M·VI·3
M·VII·3
M·I·4
M·II·4
M·III·4
M·IV·4
M·V·4
M·VI·4
M·VII·4
M·I·5
M·II·5
M·III·5
M·IV·5
M·V·5
M·VI·5
M·VII·5
M·I·6
M·II·6
M·III·6
M·IV·6
M·V·6
M·VI·6
M·VII·6
800°
720°
650°
560°
490°
420°
360°
265°
CONSTANT PRESSURE, Pa x 10⁵
CONSTANT TEMPERATURE, DEG. C
CONSTANT ENTROPY, kJ/kg K
1000
190
40
8.0
1.6
0.32
.065
608
696
764
840
916
992
1068
ENTHALPY, kJ/kg
4200
4000
3800
3600
3400
3200
3000

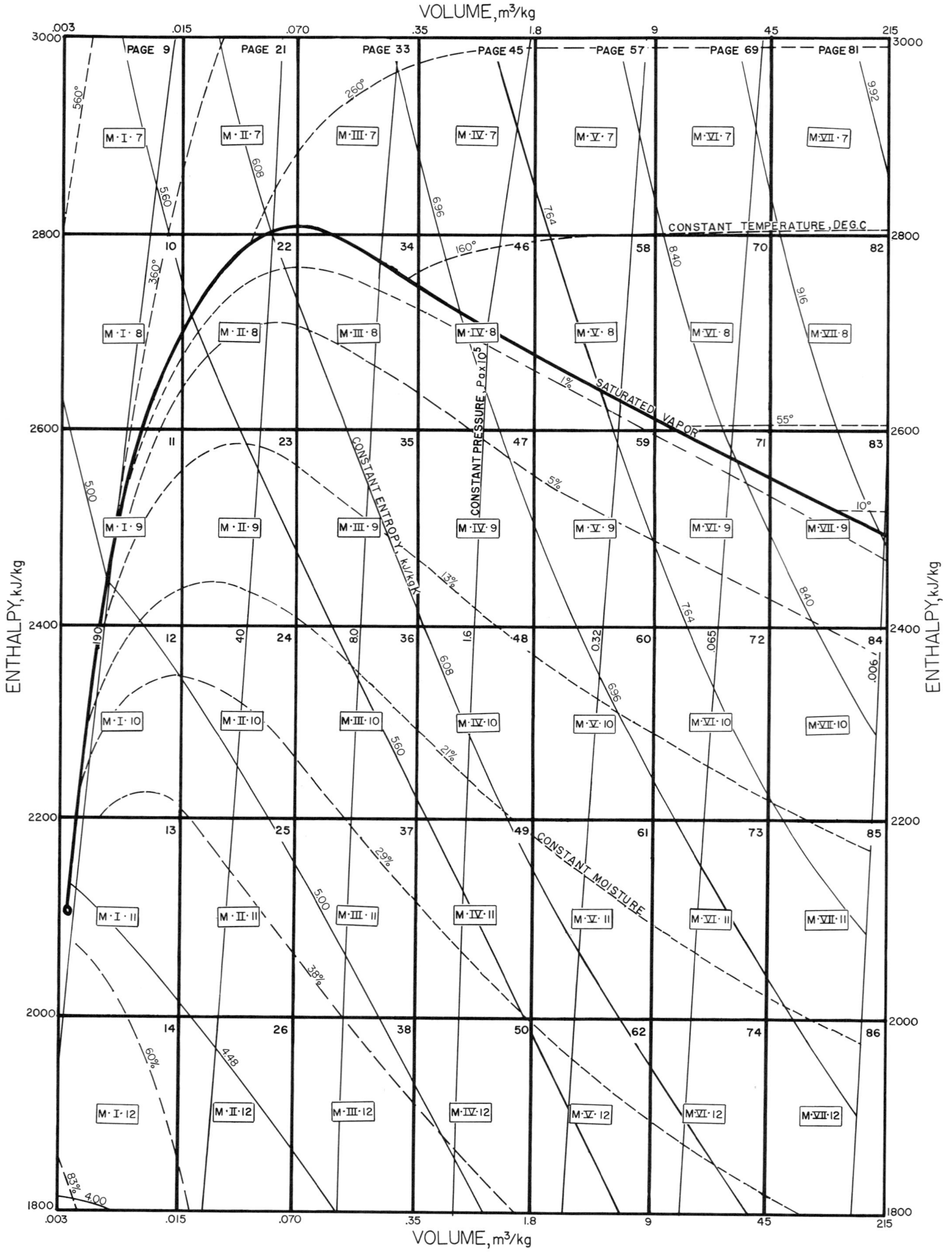
VOLUME, m³/kg
.003
.015
.070
.35
1.8
9
45
215
ENTHALPY, kJ/kg
3000
2800
2600
2400
2200
2000
1800
PAGE 9
PAGE 21
PAGE 33
PAGE 45
PAGE 57
PAGE 69
PAGE 81
M·I·7
M·II·7
M·III·7
M·IV·7
M·V·7
M·VI·7
M·VII·7
M·I·8
M·II·8
M·III·8
M·IV·8
M·V·8
M·VI·8
M·VII·8
M·I·9
M·II·9
M·III·9
M·IV·9
M·V·9
M·VI·9
M·VII·9
M·I·10
M·II·10
M·III·10
M·IV·10
M·V·10
M·VI·10
M·VII·10
M·I·11
M·II·11
M·III·11
M·IV·11
M·V·11
M·VI·11
M·VII·11
M·I·12
M·II·12
M·III·12
M·IV·12
M·V·12
M·VI·12
M·VII·12
10
22
34
46
58
70
82
11
23
35
47
59
71
83
12
24
36
48
60
72
84
13
25
37
49
61
73
85
14
26
38
50
62
74
86
CONSTANT TEMPERATURE, DEG.C
560°
360°
260°
160°
55°
10°
SATURATED VAPOR
CONSTANT PRESSURE, Pa x 10⁵
190
40
8.0
1.6
0.32
.065
.006
CONSTANT ENTROPY, kJ/kg K
560
608
696
764
840
916
992
500
448
400
CONSTANT MOISTURE
1%
5%
13%
21%
29%
38%
60%
83%

COMPRESSED LIQUID
S I (Metric) units

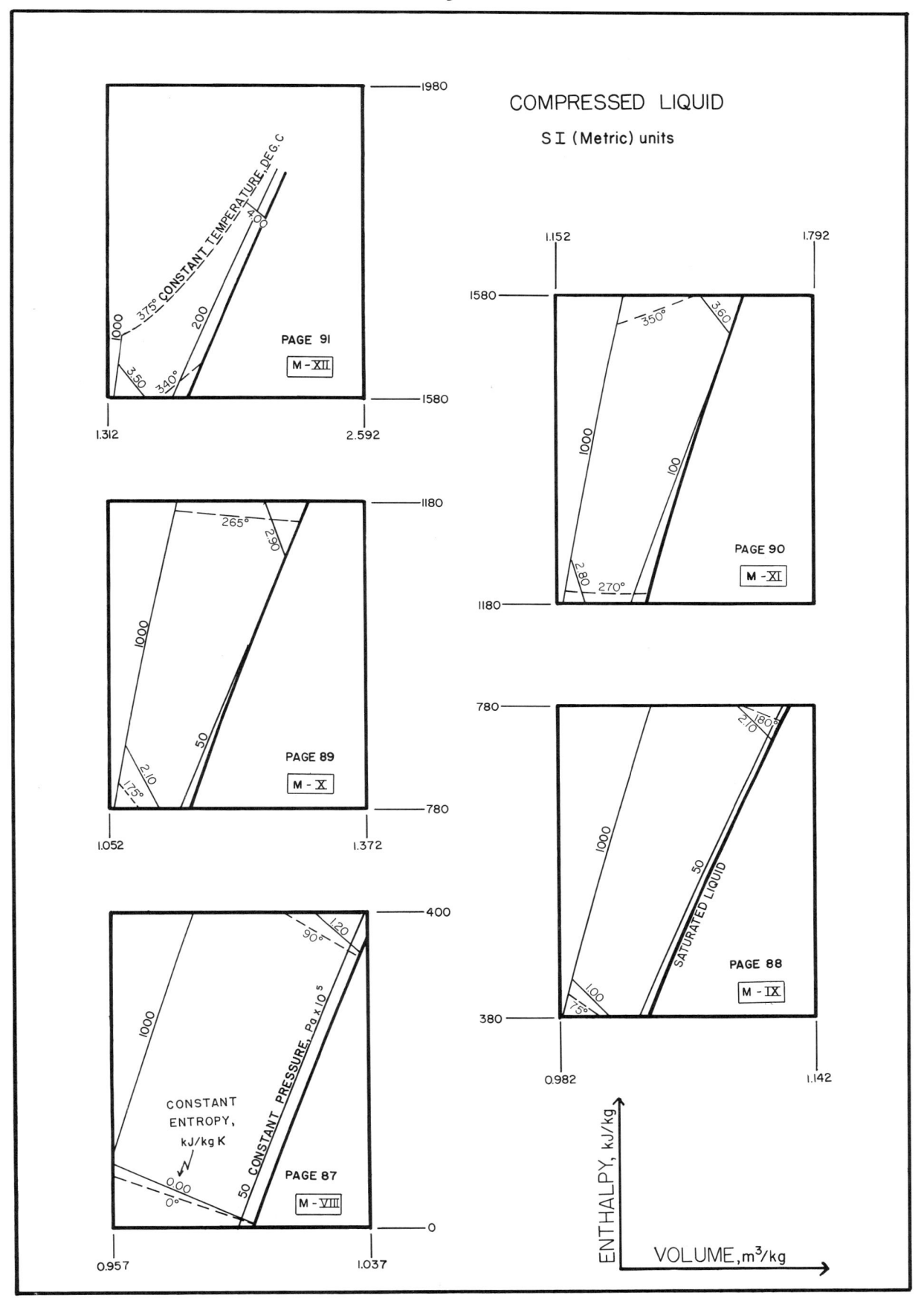

PLATE M-I-2

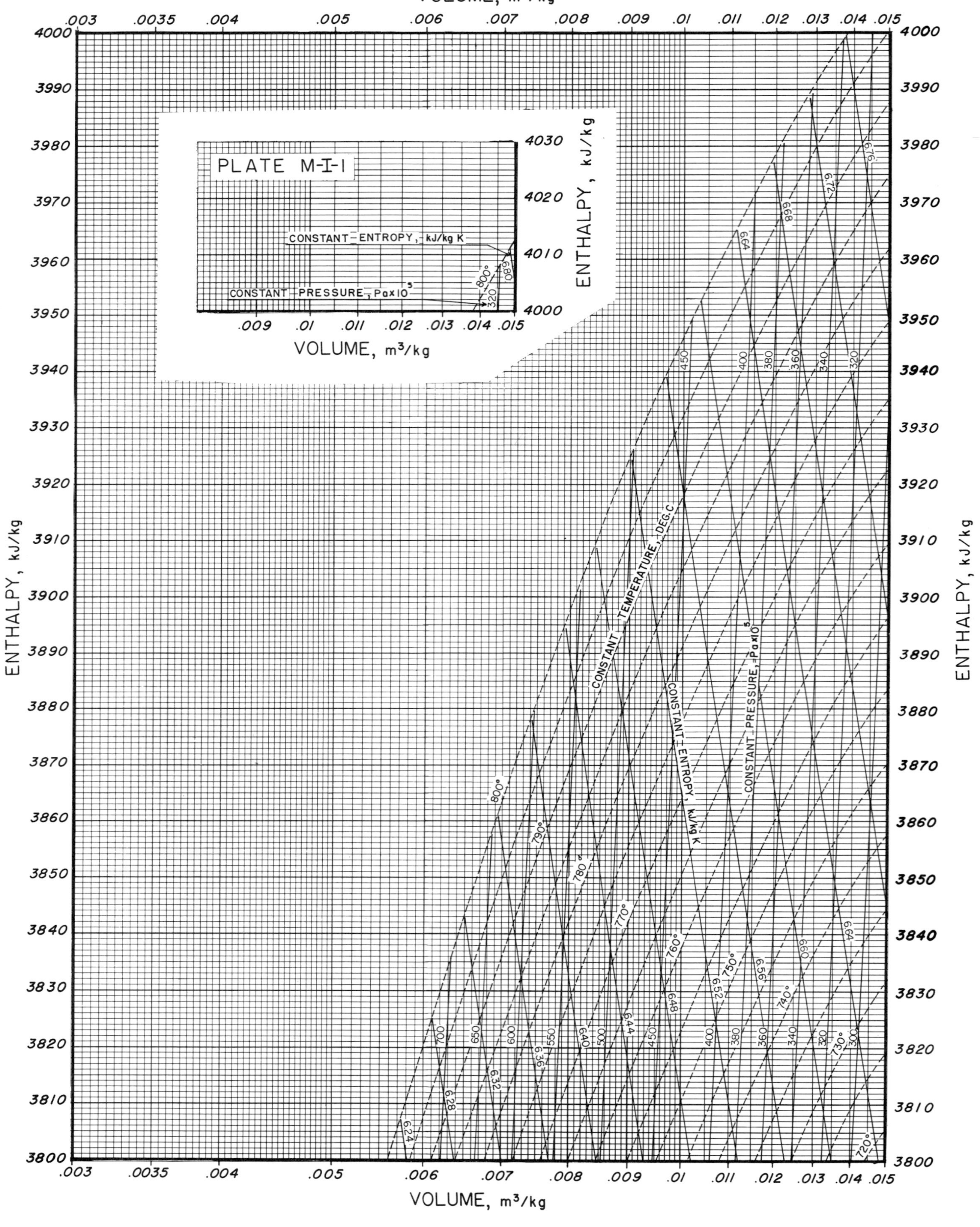
VOLUME, m³/kg
.003 .0035 .004 .005 .006 .007 .008 .009 .01 .011 .012 .013 .014 .015
ENTHALPY, kJ/kg
4000 3990 3980 3970 3960 3950 3940 3930 3920 3910 3900 3890 3880 3870 3860 3850 3840 3830 3820 3810 3800
PLATE M-I-1
CONSTANT ENTROPY, kJ/kg K
CONSTANT PRESSURE, Pa x 10⁵
4030 4020 4010 4000
.009 .01 .011 .012 .013 .014 .015
CONSTANT TEMPERATURE, DEG. C
CONSTANT ENTROPY, kJ/kg K
CONSTANT PRESSURE, Pa x 10⁵
800° 790° 780° 770° 760° 750° 740° 730° 720°
6.24 6.28 6.32 6.36 6.40 6.44 6.48 6.52 6.56 6.60 6.64 6.68 6.72 6.76
700 650 600 550 500 450 400 380 360 340 320 300

PLATE M-I-3

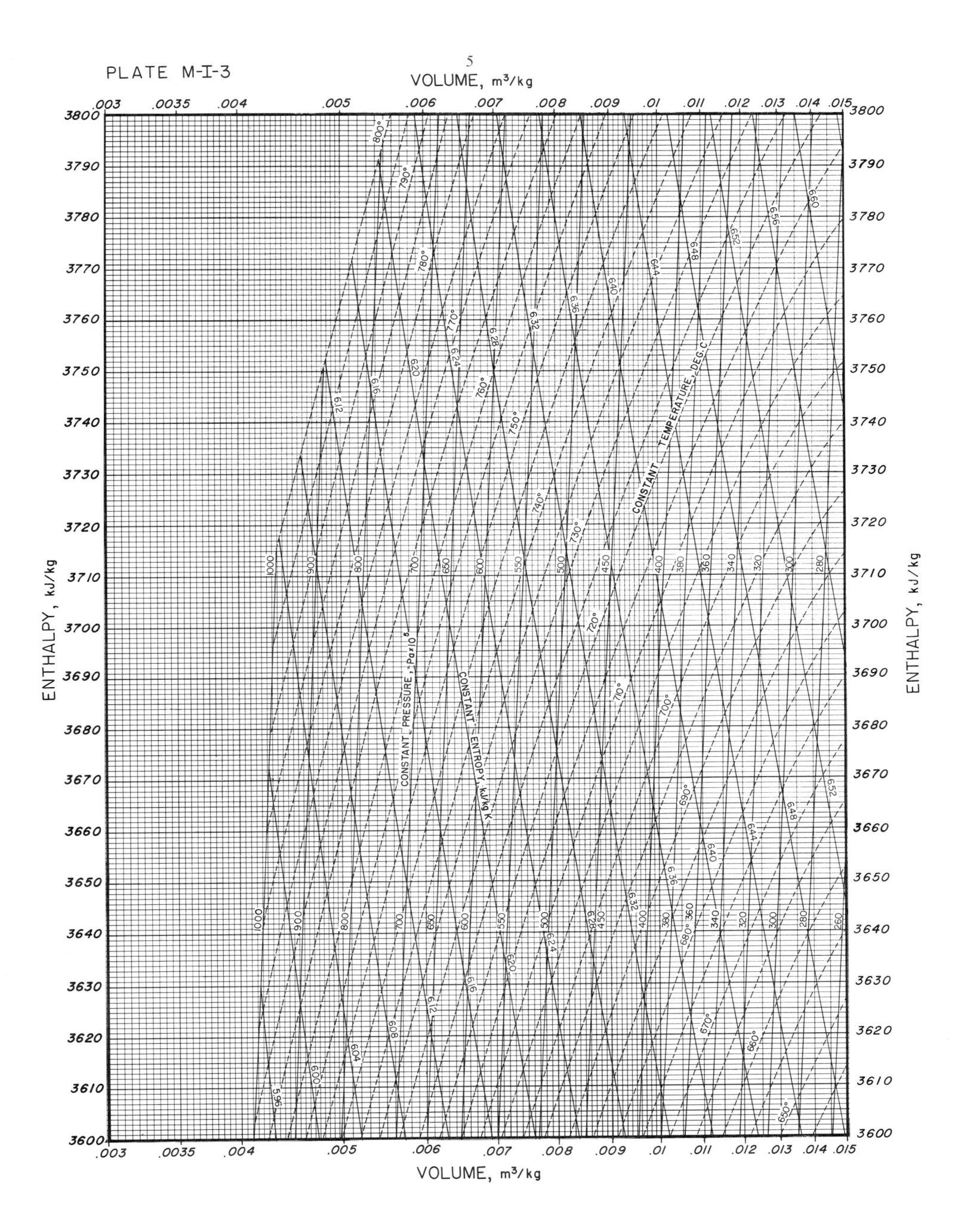

PLATE M-I-4

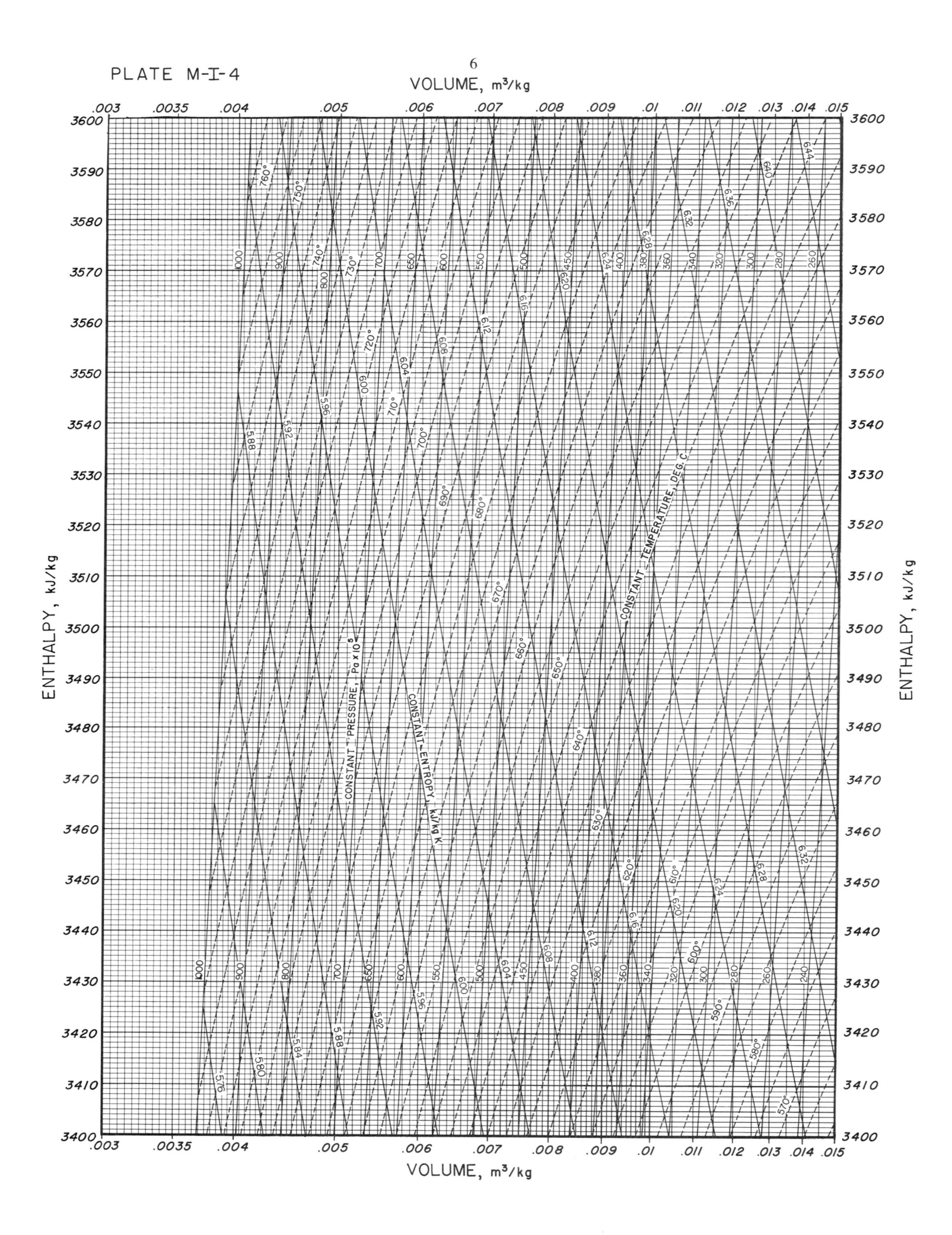

PLATE M-I-5

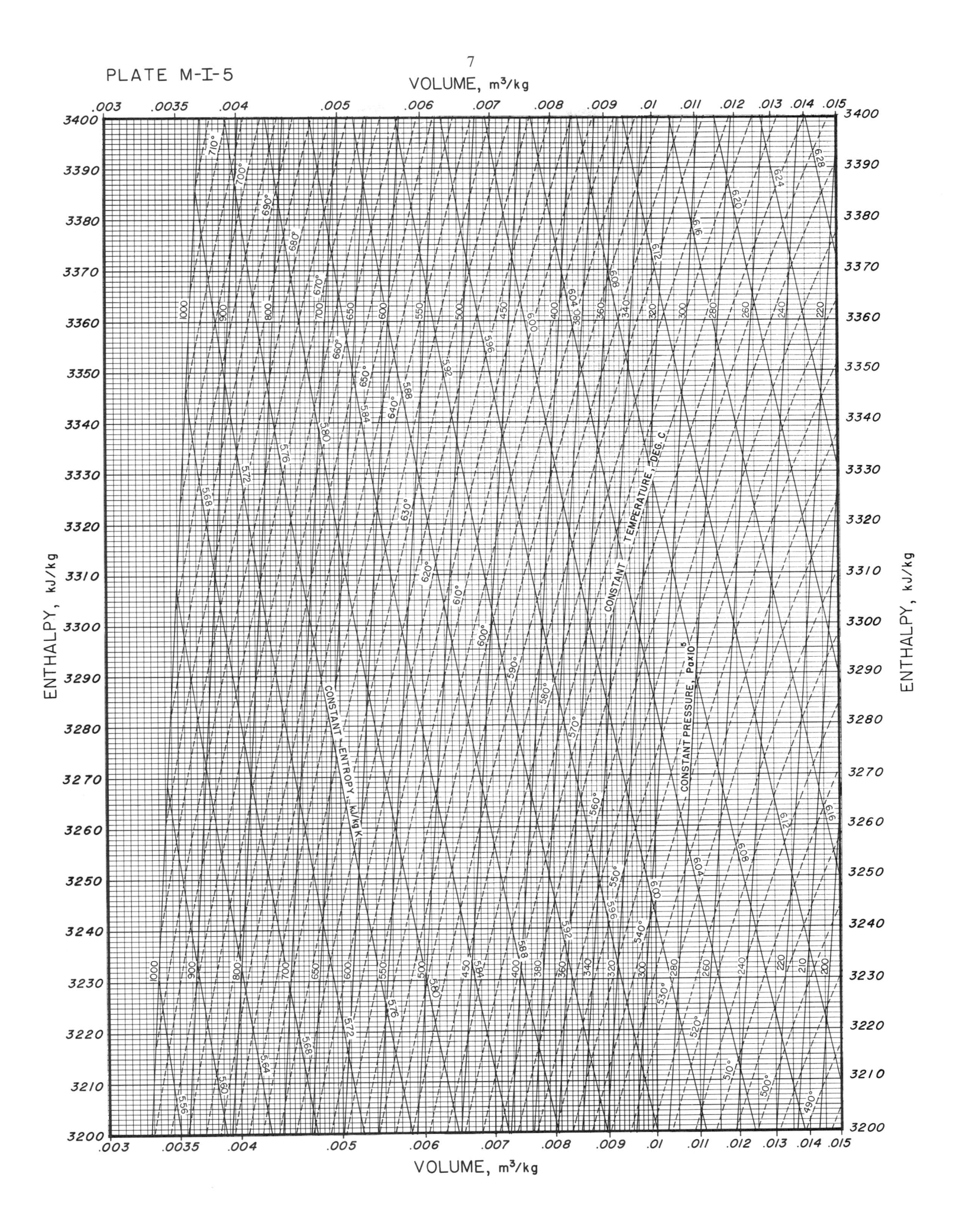

PLATE M-I-6

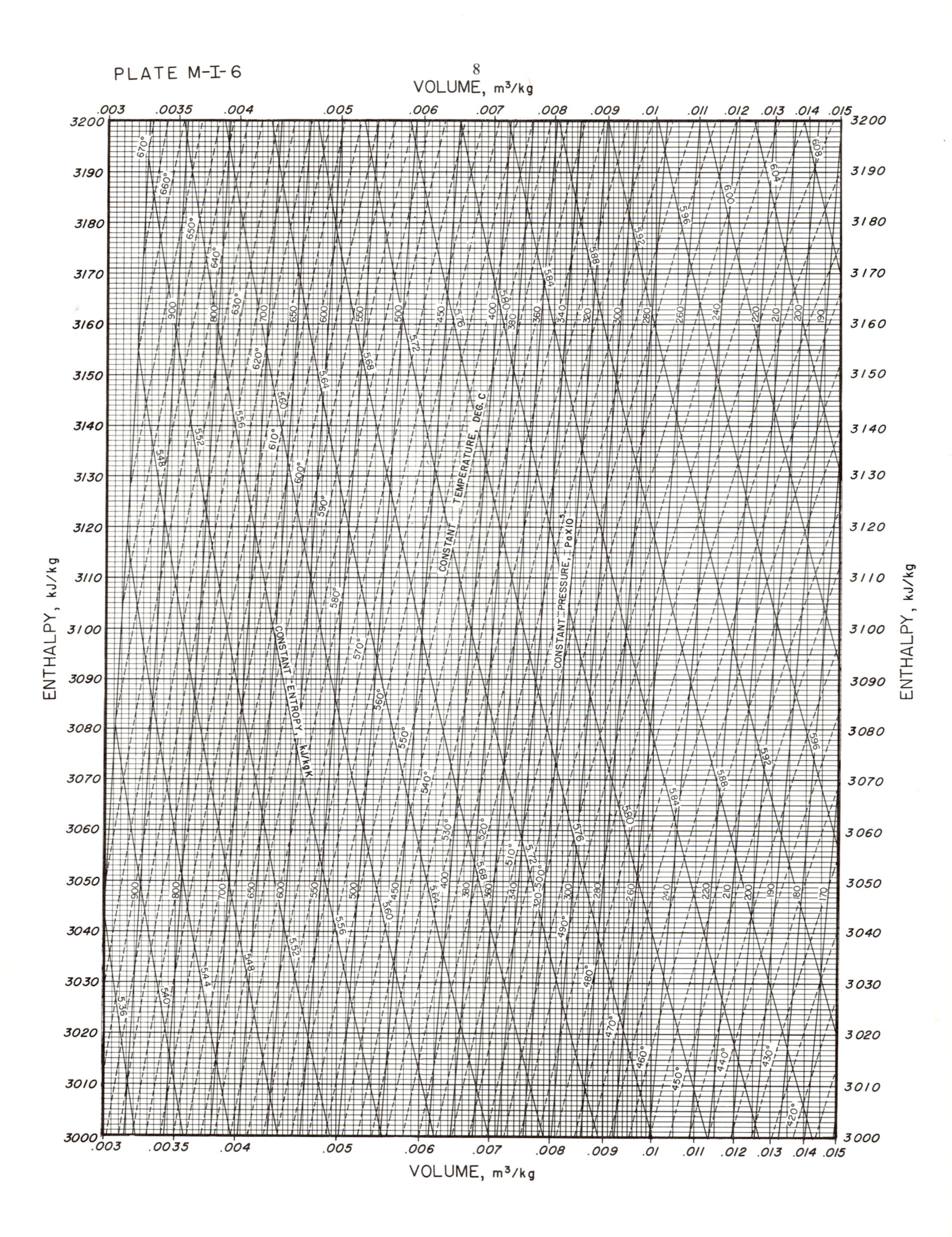

PLATE M-I-7

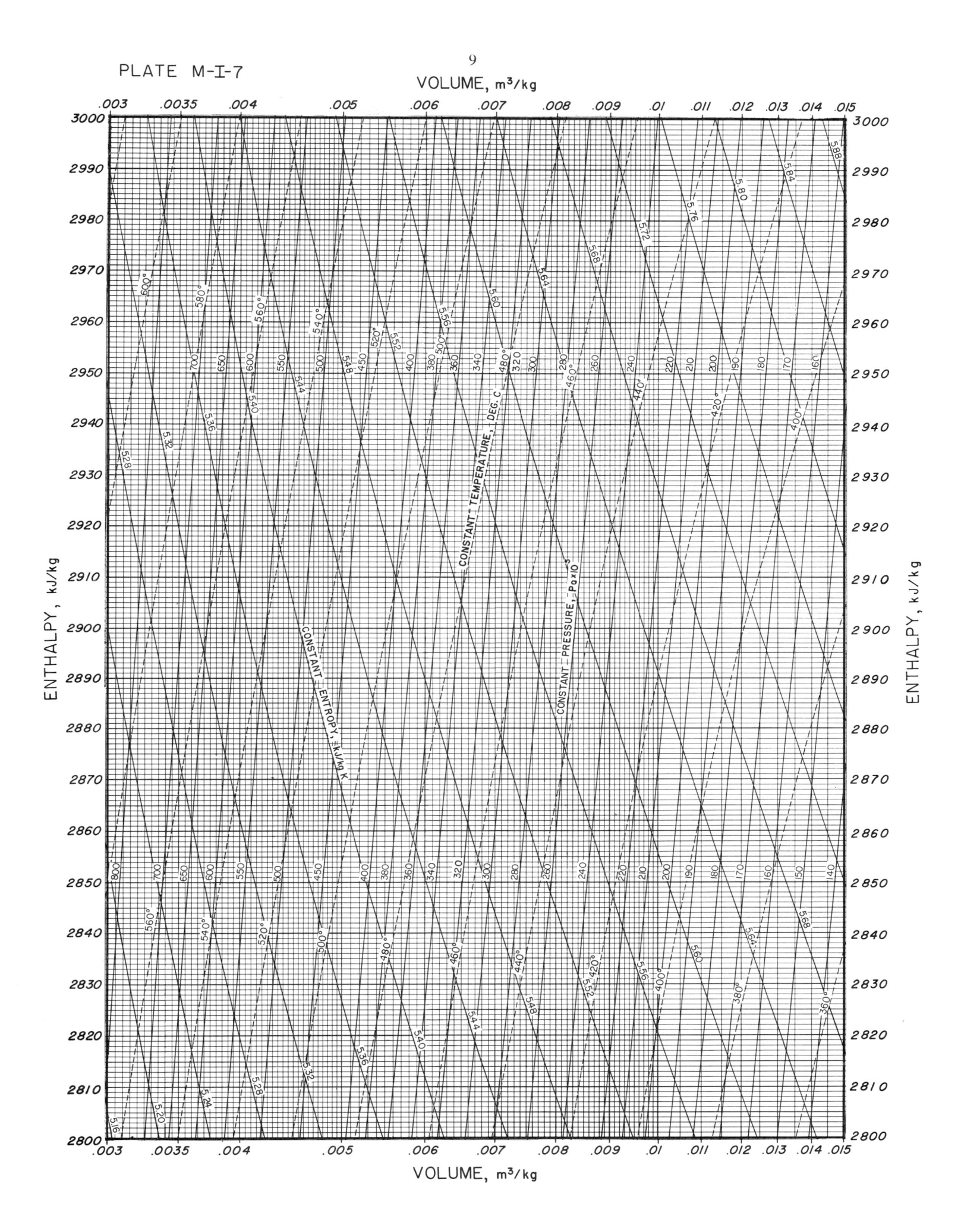

PLATE M-I-8

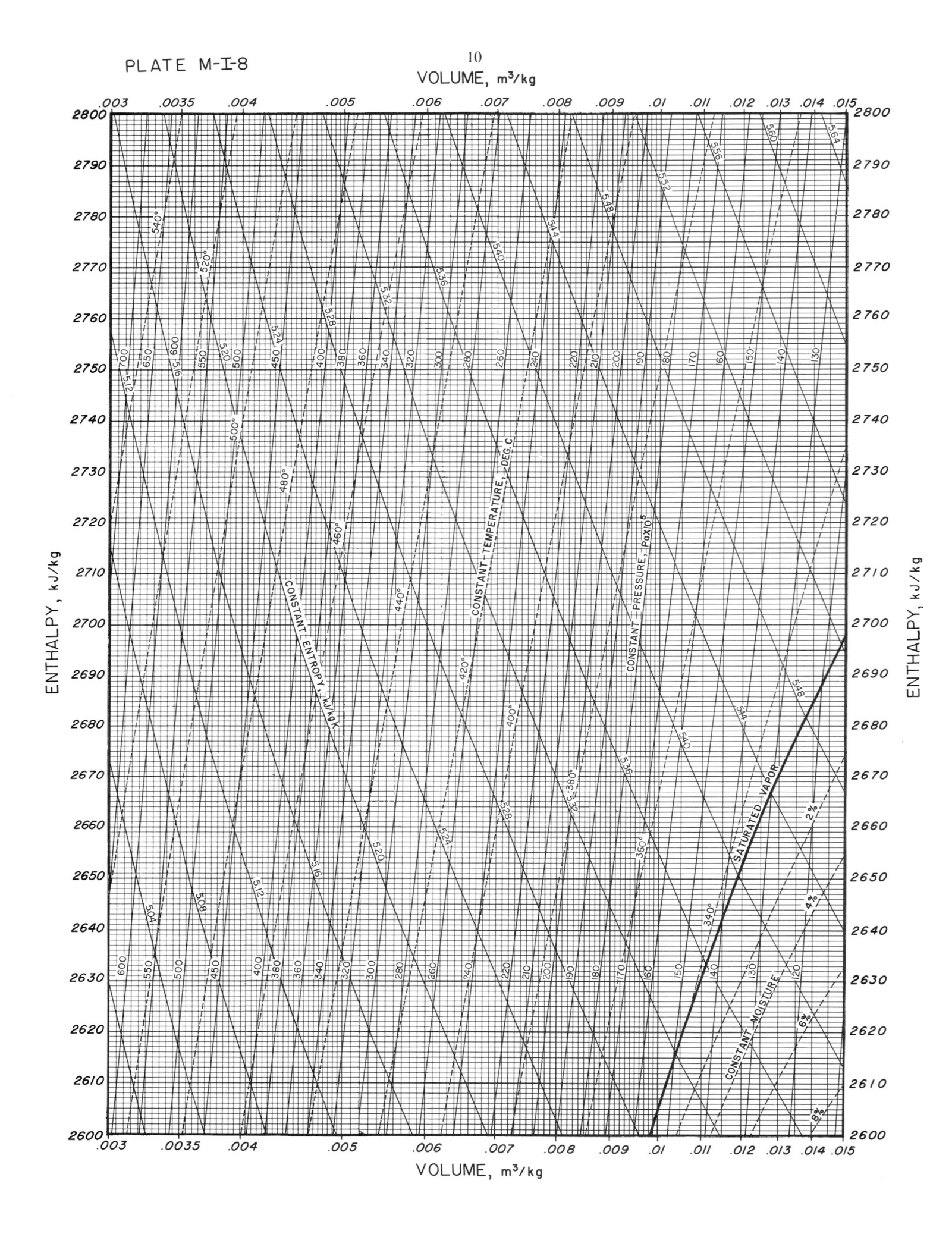

PLATE M-I-9

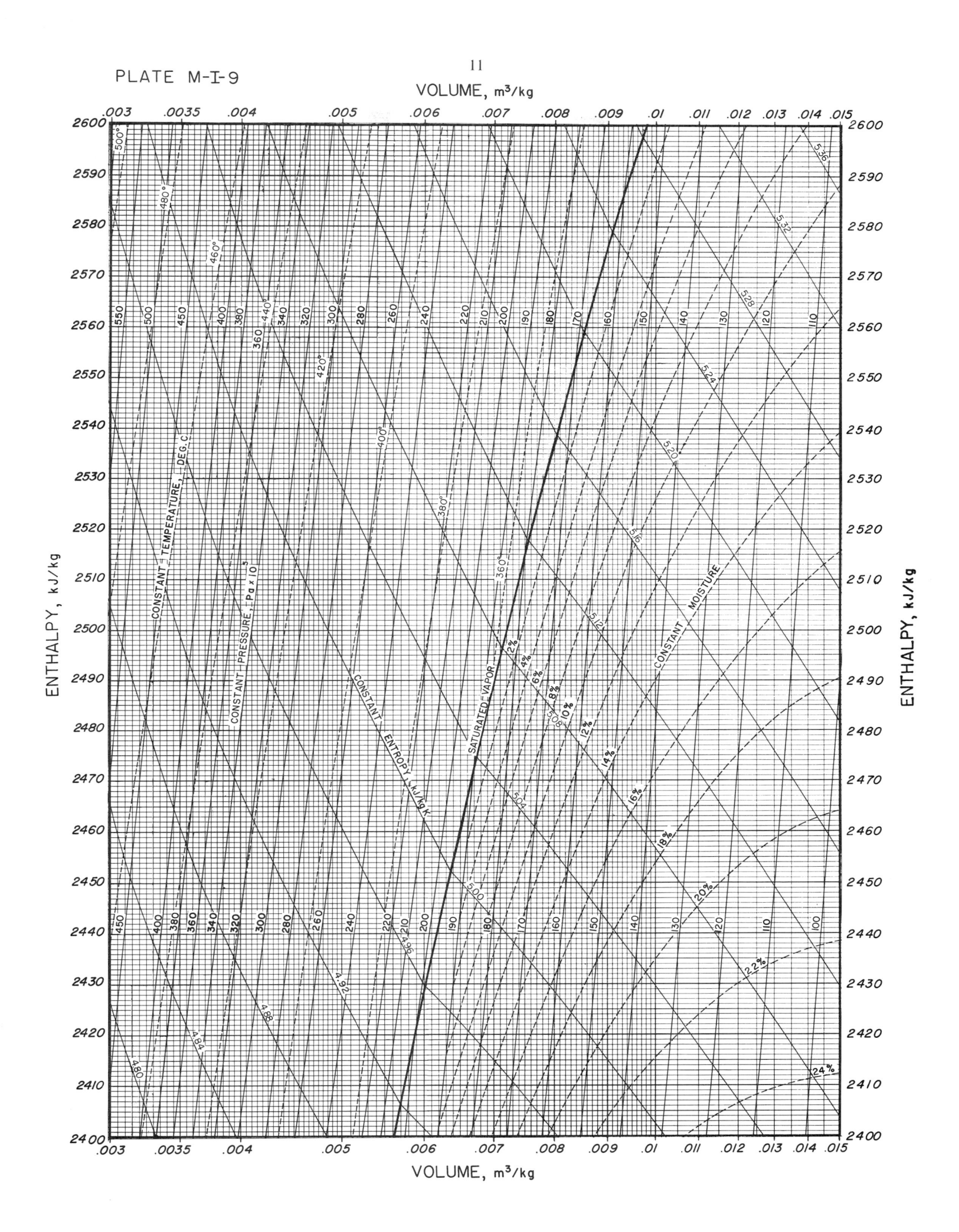

PLATE M-I-10

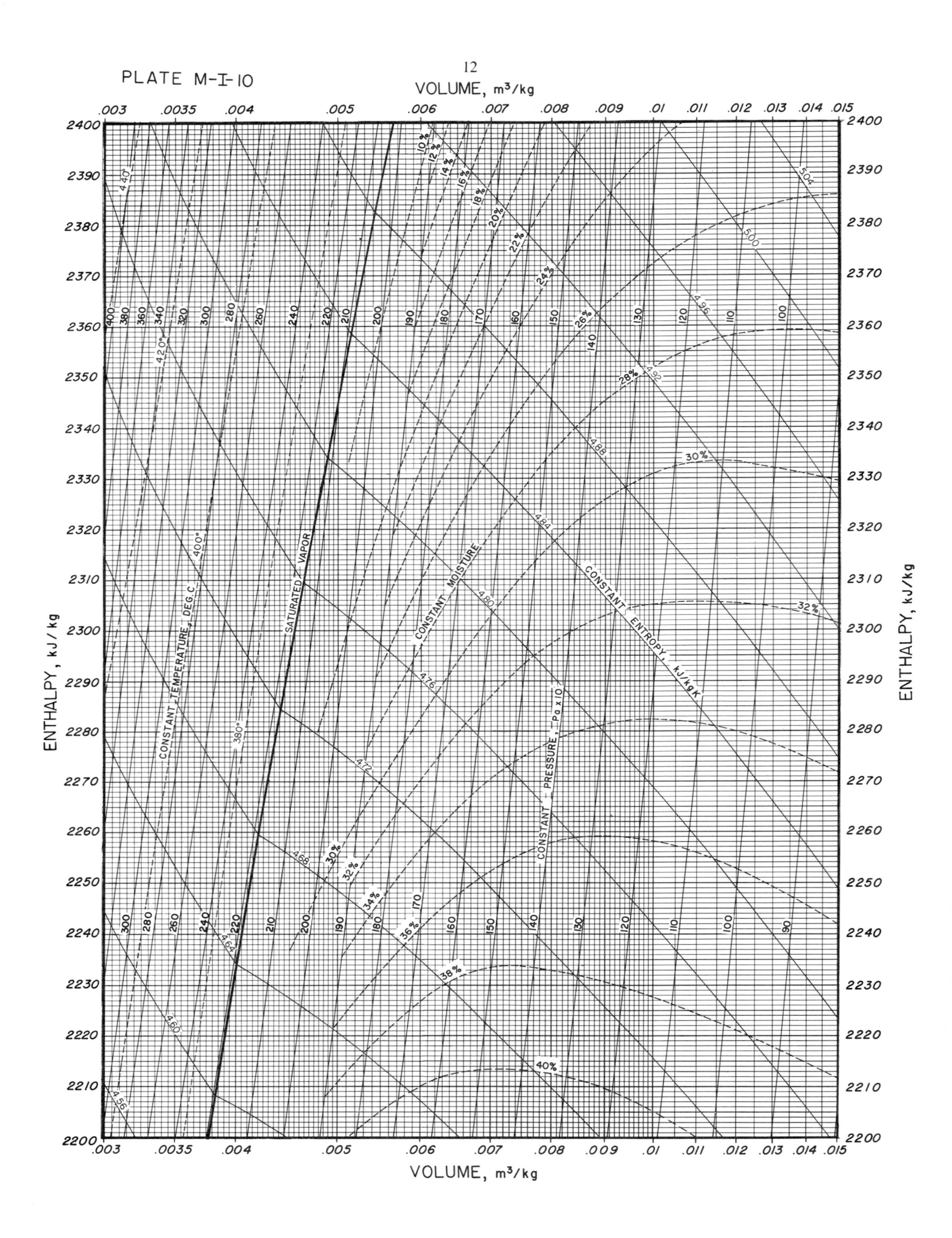

PLATE M-I-11

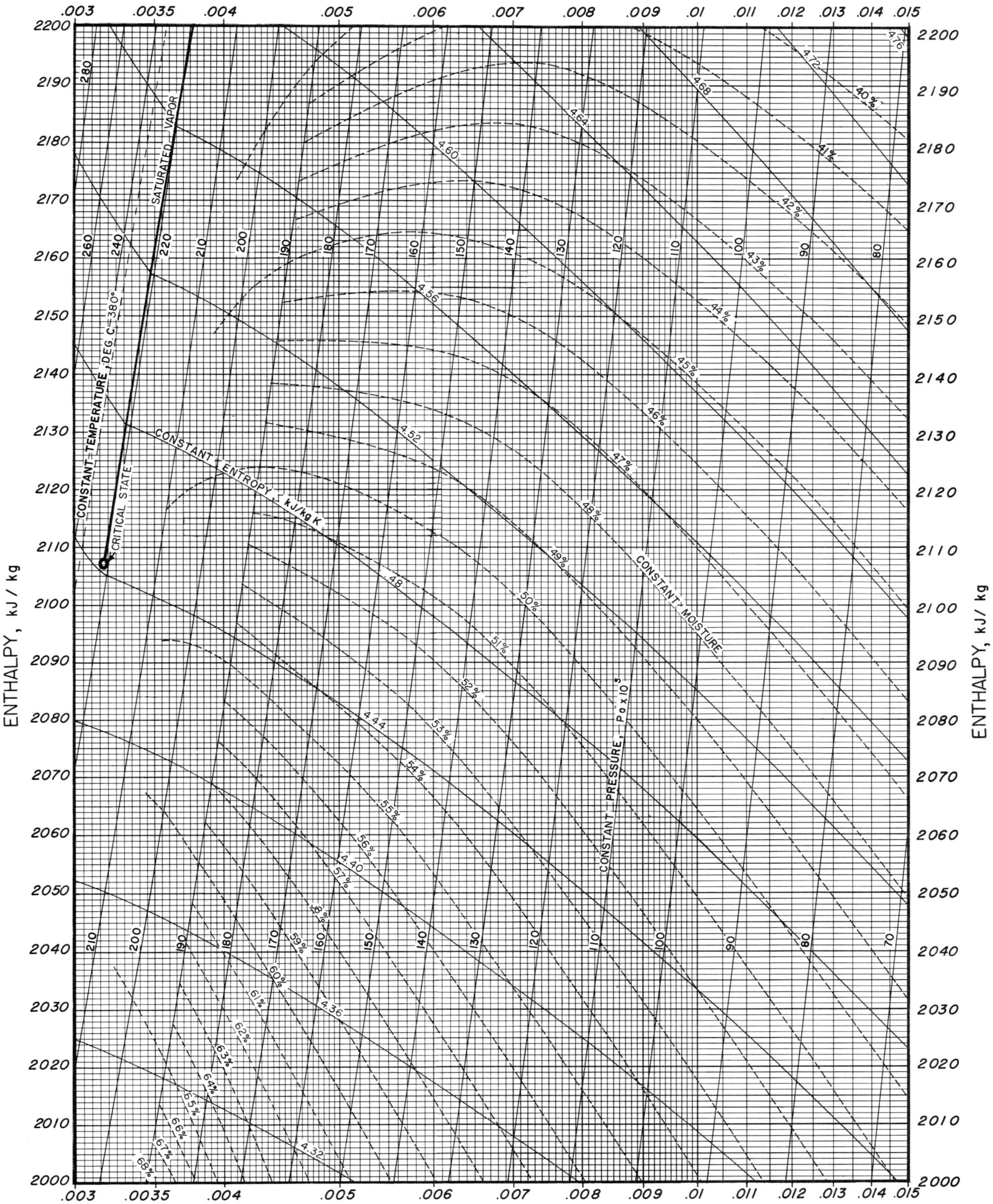
VOLUME, m³/kg
ENTHALPY, kJ / kg
ENTHALPY, kJ/ kg
SATURATED VAPOR
CONSTANT-TEMPERATURE, DEG. C=380°
CRITICAL STATE
CONSTANT ENTROPY, kJ/kg K
CONSTANT MOISTURE
CONSTANT PRESSURE, Pa x 10⁵
.003
.0035
.004
.005
.006
.007
.008
.009
.01
.011
.012
.013
.014
.015
2000
2010
2020
2030
2040
2050
2060
2070
2080
2090
2100
2110
2120
2130
2140
2150
2160
2170
2180
2190
2200

PLATE M-1-12

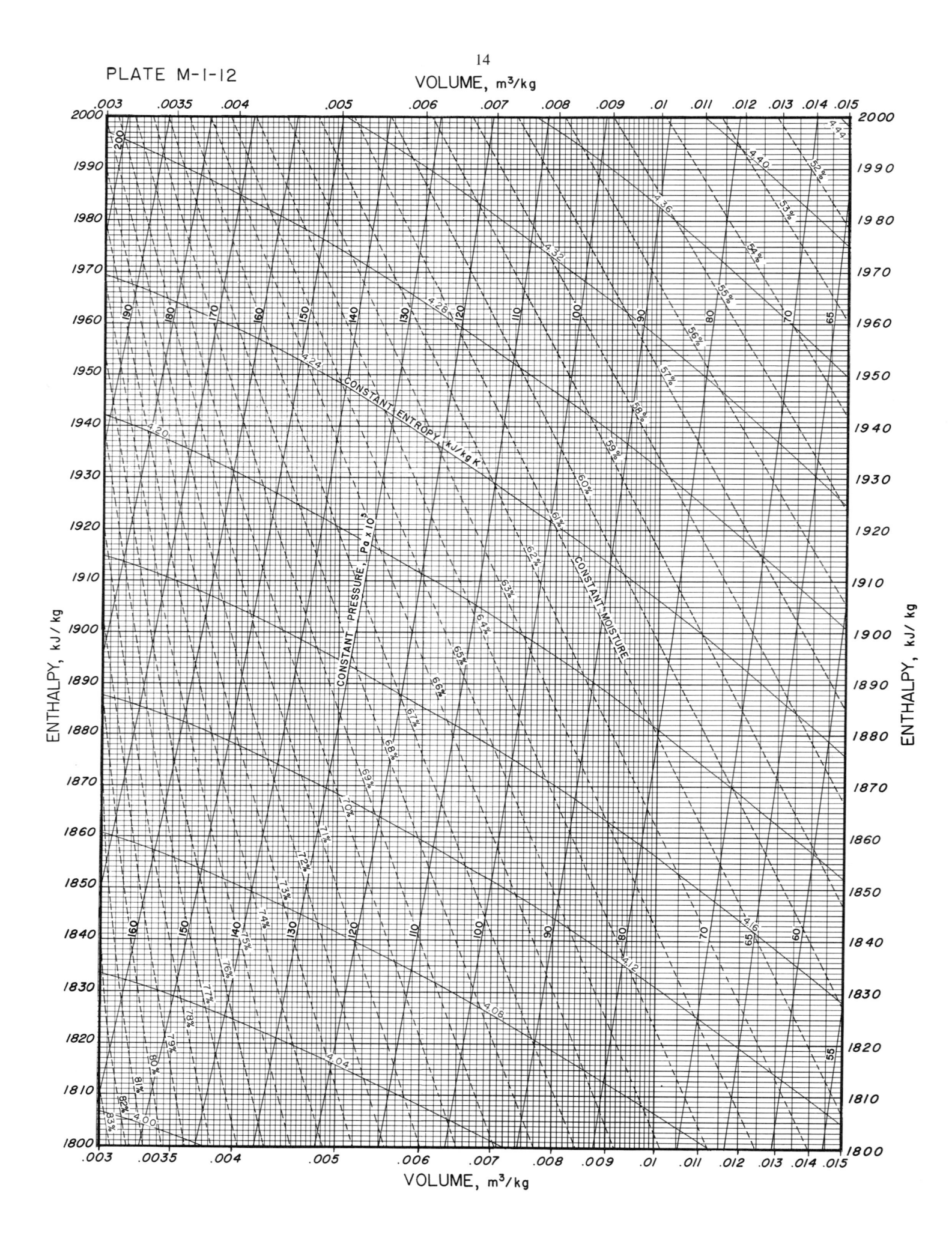

PLATE M-II-1

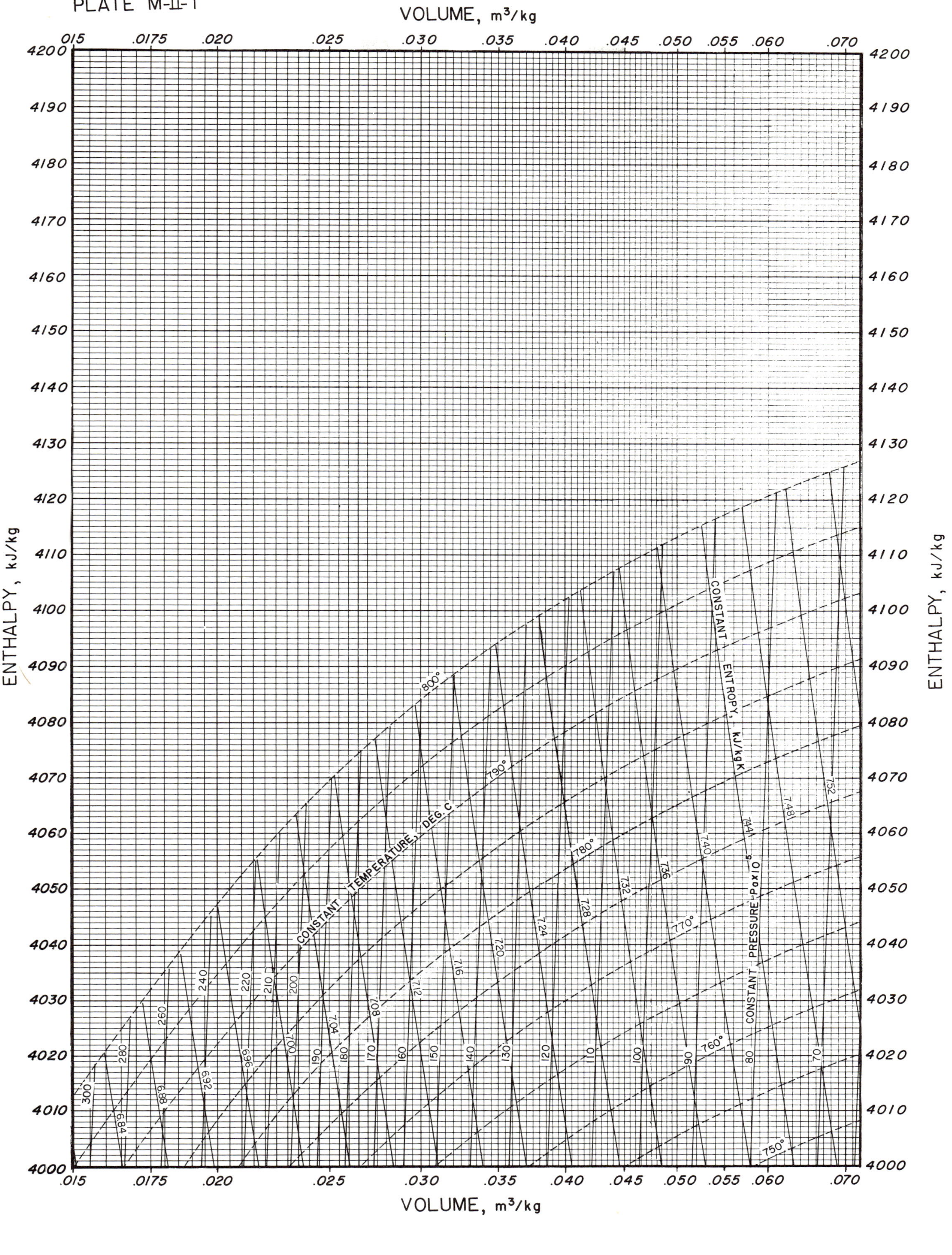

VOLUME, m³/kg
ENTHALPY, kJ/kg
CONSTANT TEMPERATURE, DEG. C
CONSTANT ENTROPY, kJ/kgK
CONSTANT PRESSURE-Pa×10⁵

PLATE M-II-2

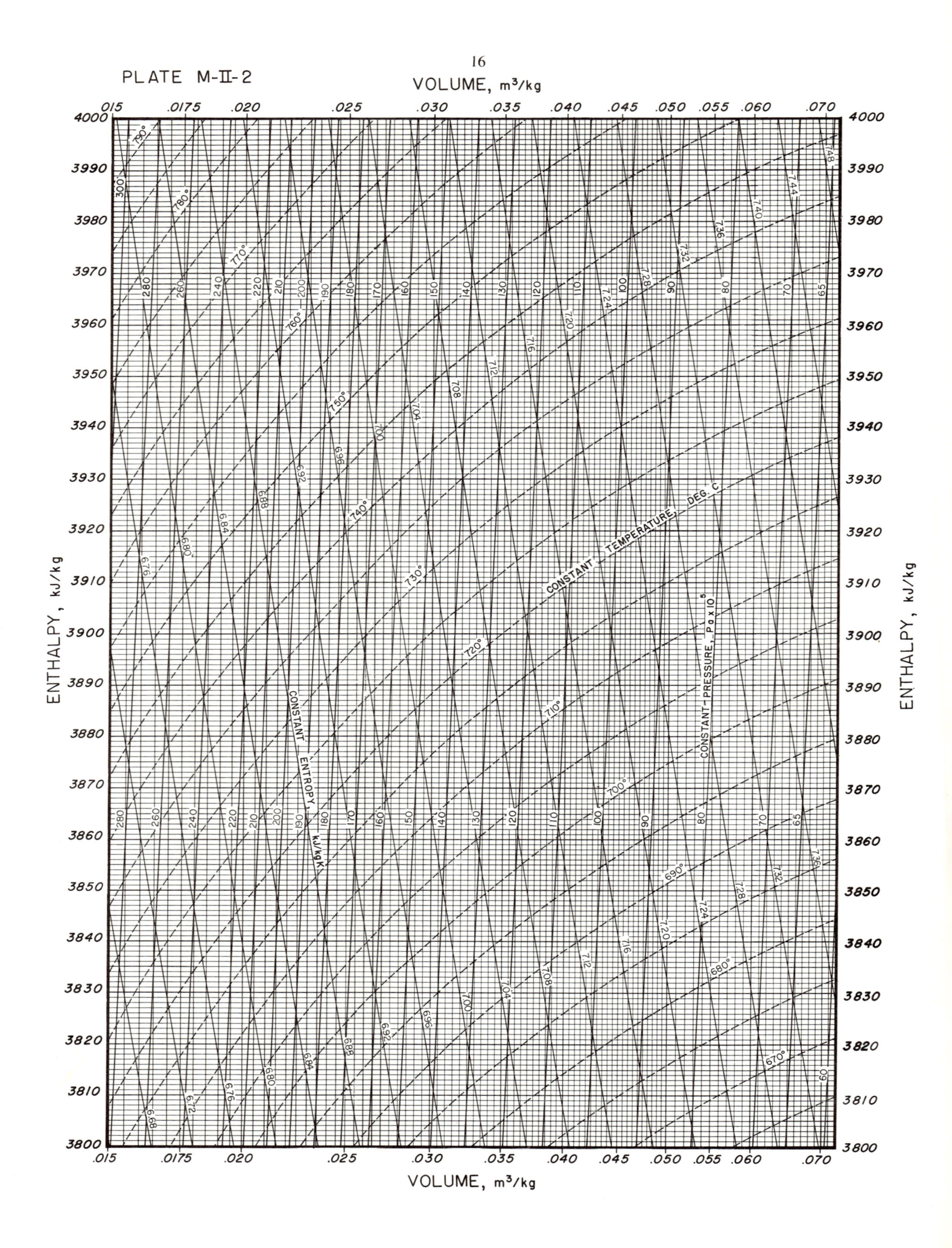

PLATE M-II-3

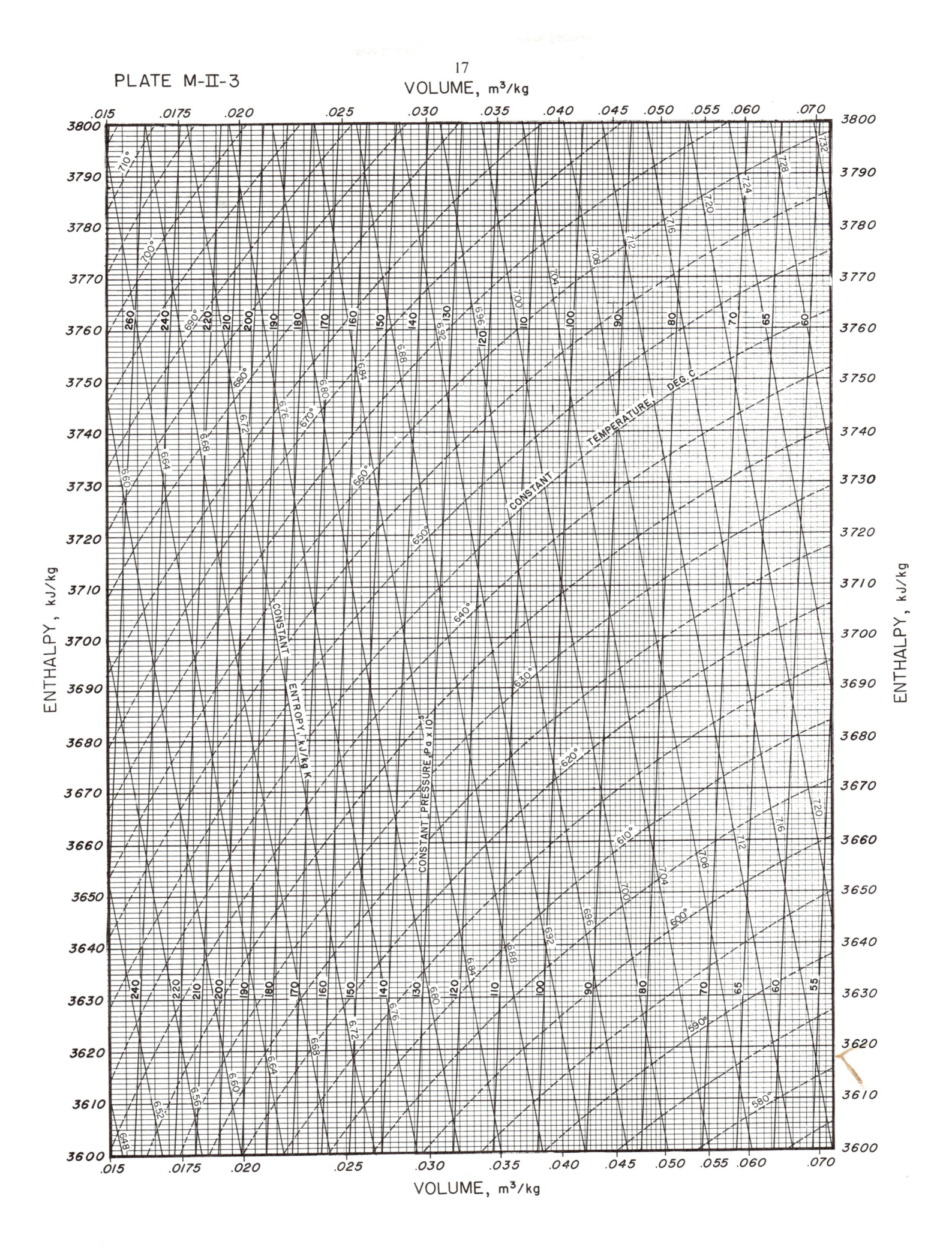

PLATE M-II-4

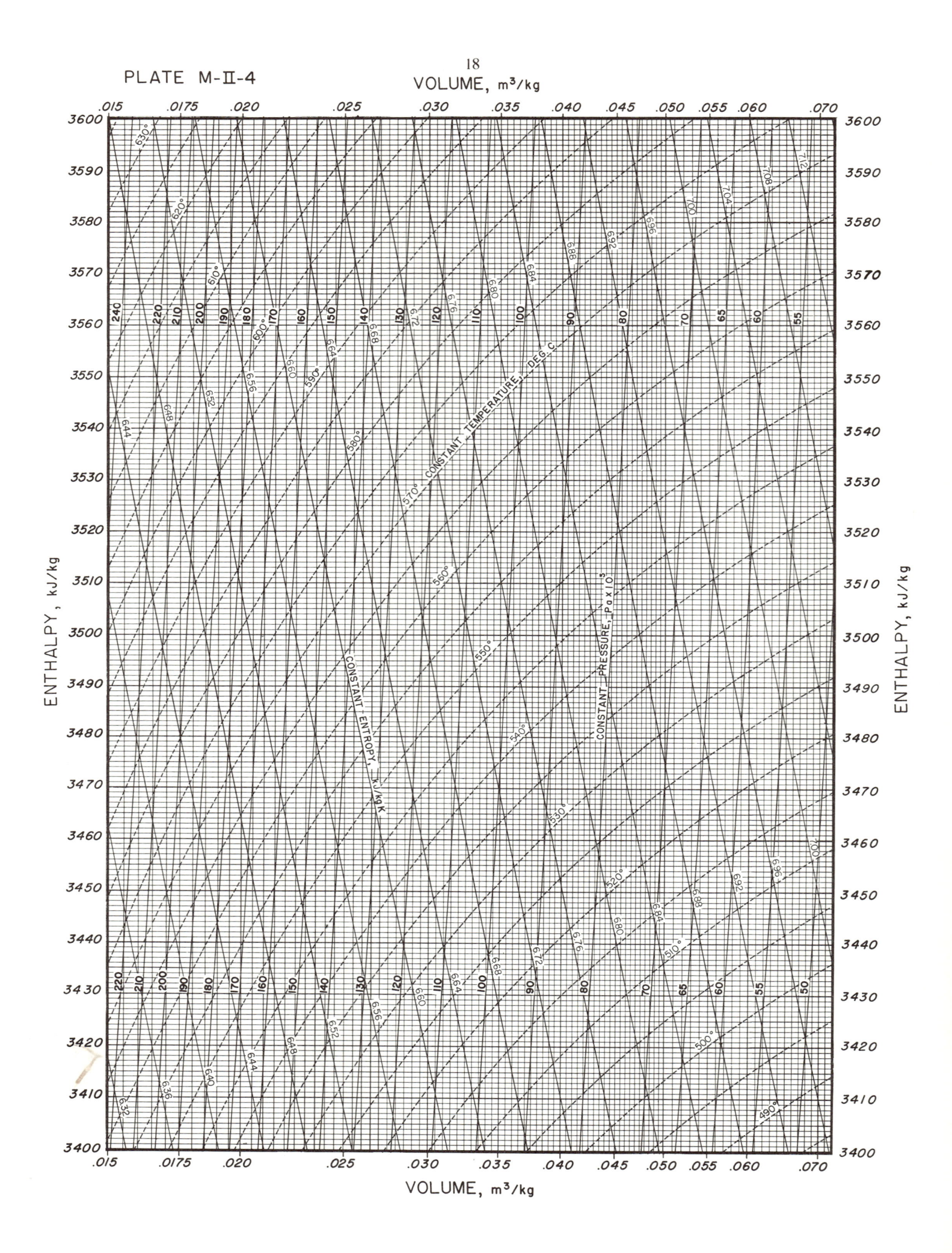

PLATE M-II-5

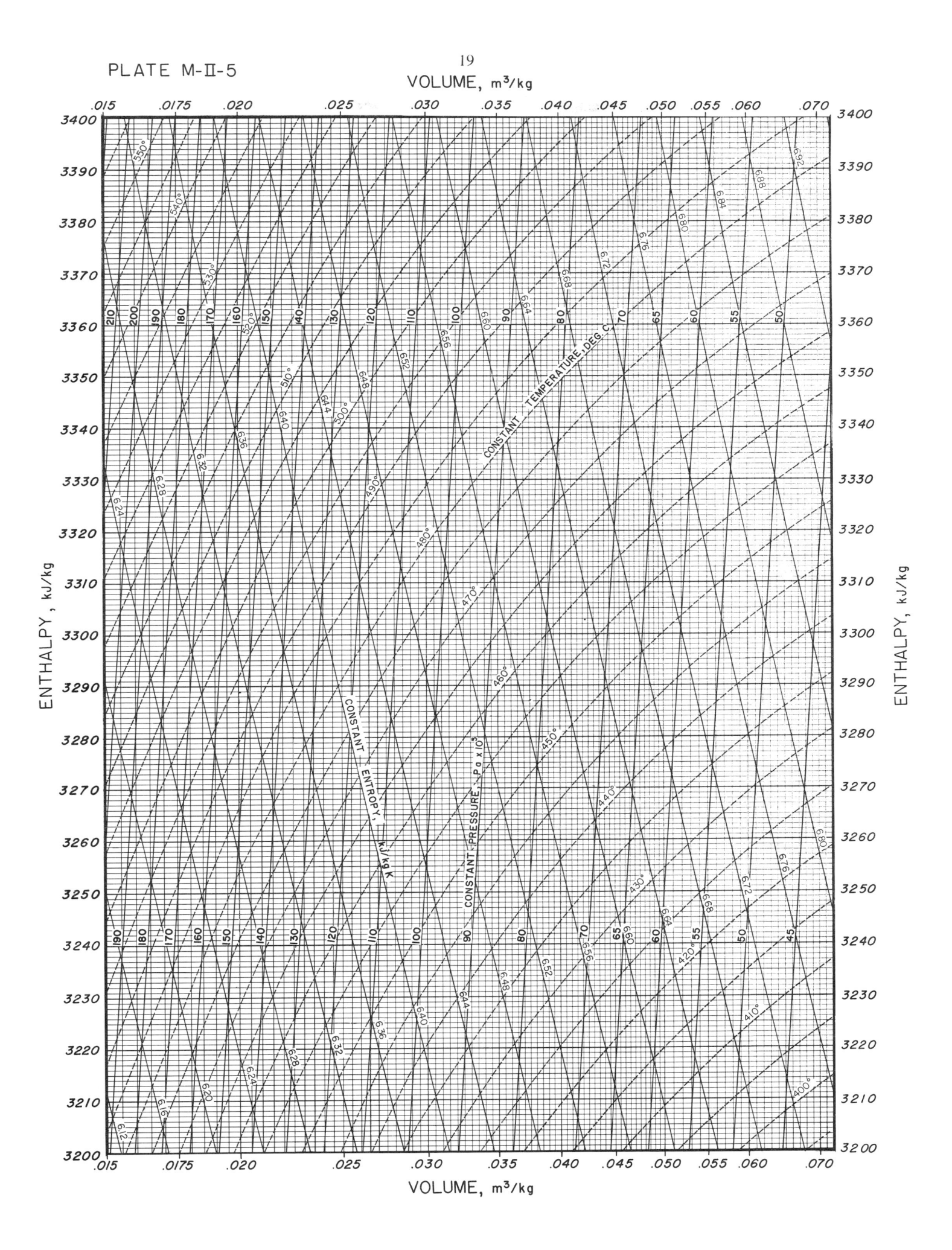

PLATE M-II-6

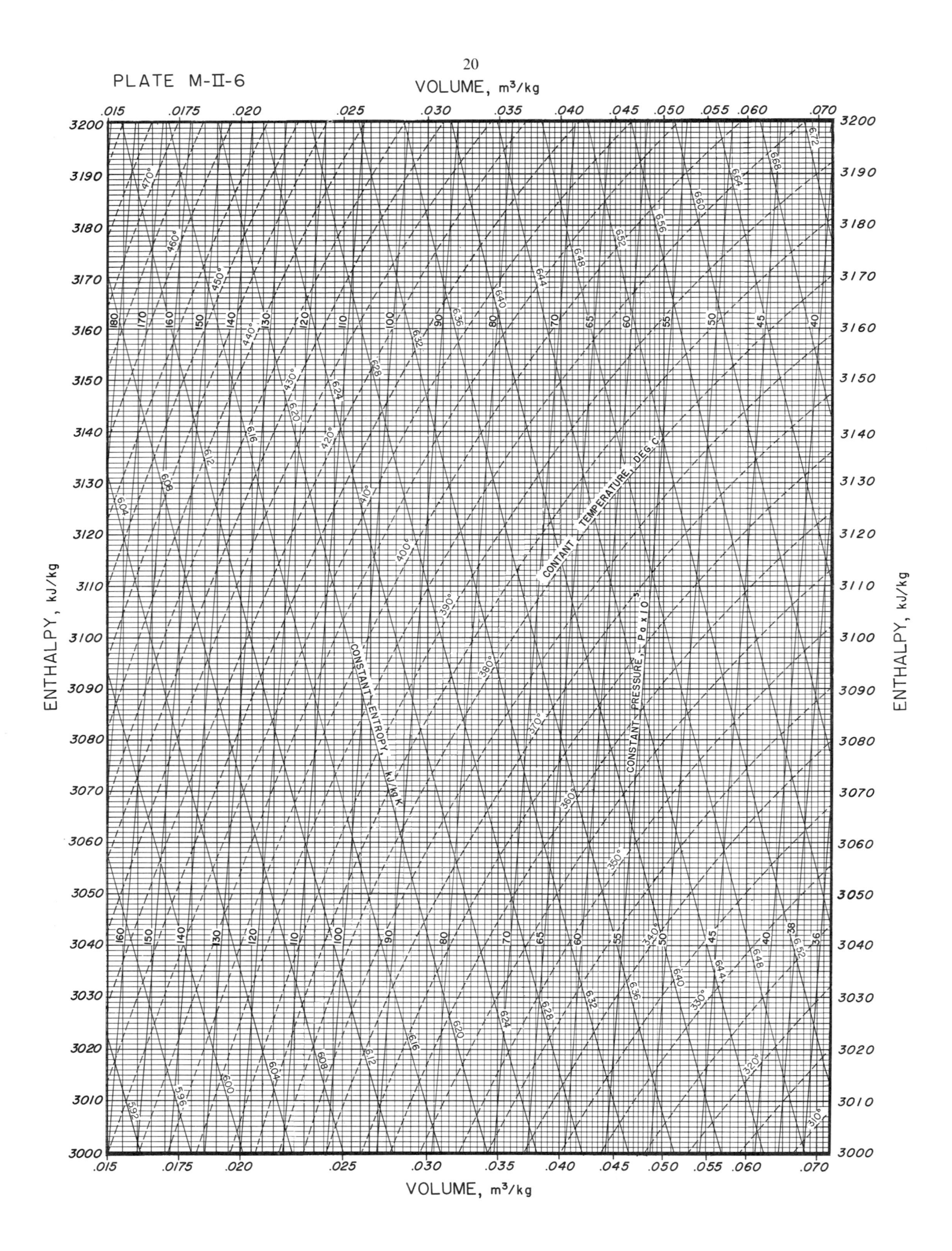

PLATE M-II-7

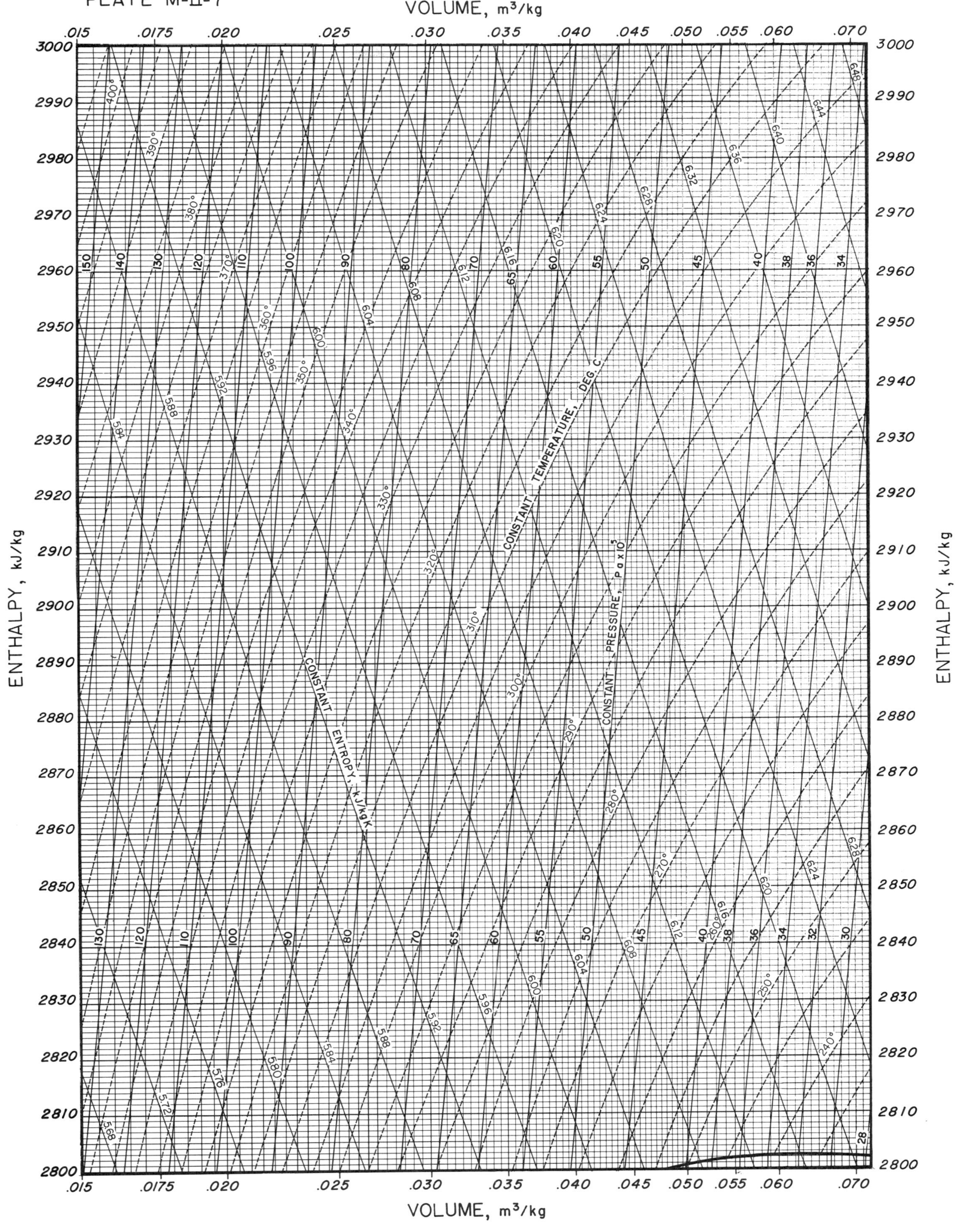
VOLUME, m³/kg
.015 .0175 .020 .025 .030 .035 .040 .045 .050 .055 .060 .070
ENTHALPY, kJ/kg
3000 2990 2980 2970 2960 2950 2940 2930 2920 2910 2900 2890 2880 2870 2860 2850 2840 2830 2820 2810 2800
CONSTANT TEMPERATURE, DEG. C
CONSTANT PRESSURE, Pa x 10⁵
CONSTANT ENTROPY, kJ/kg K

PLATE M-II-8

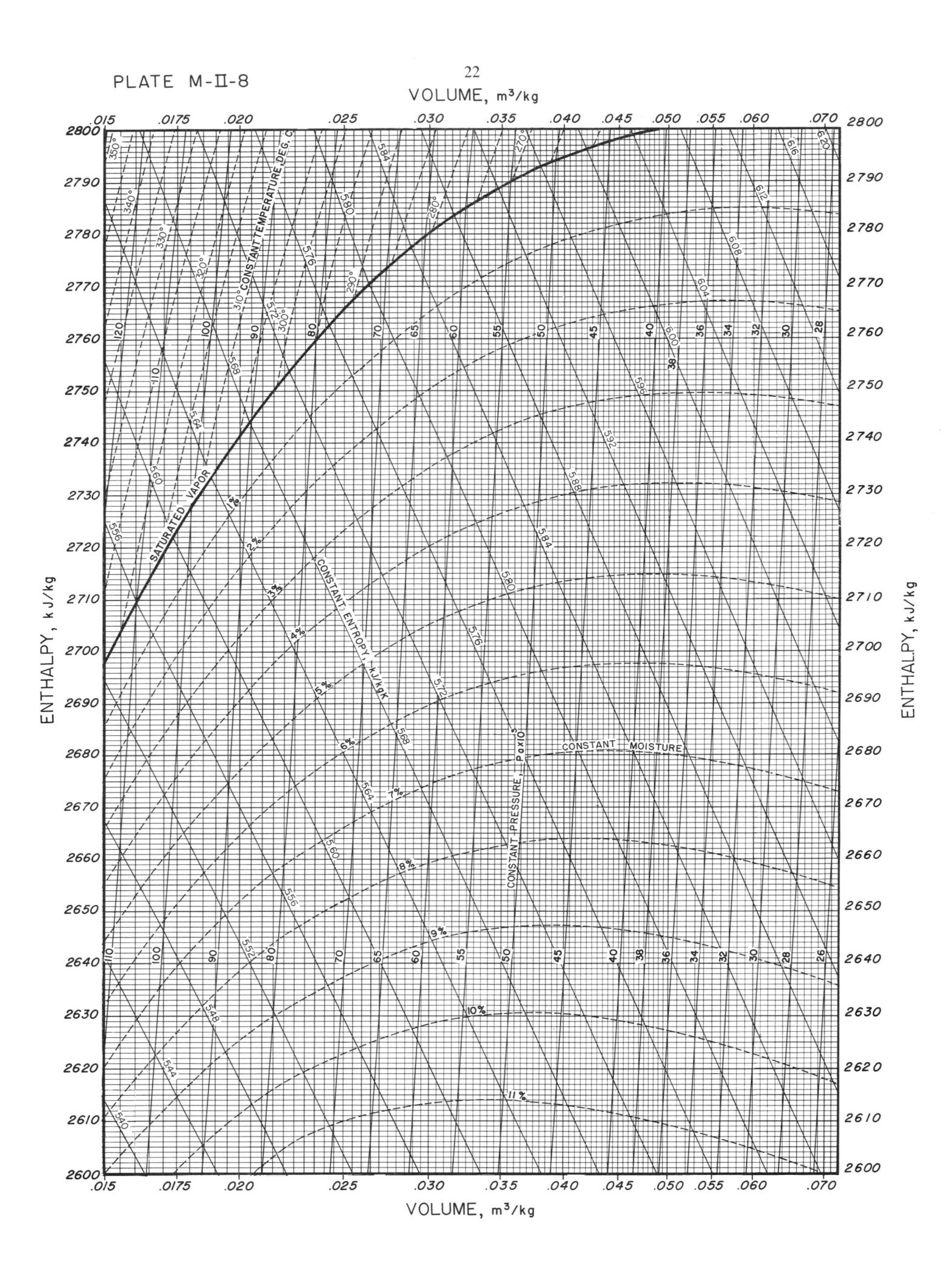

PLATE M-II-9

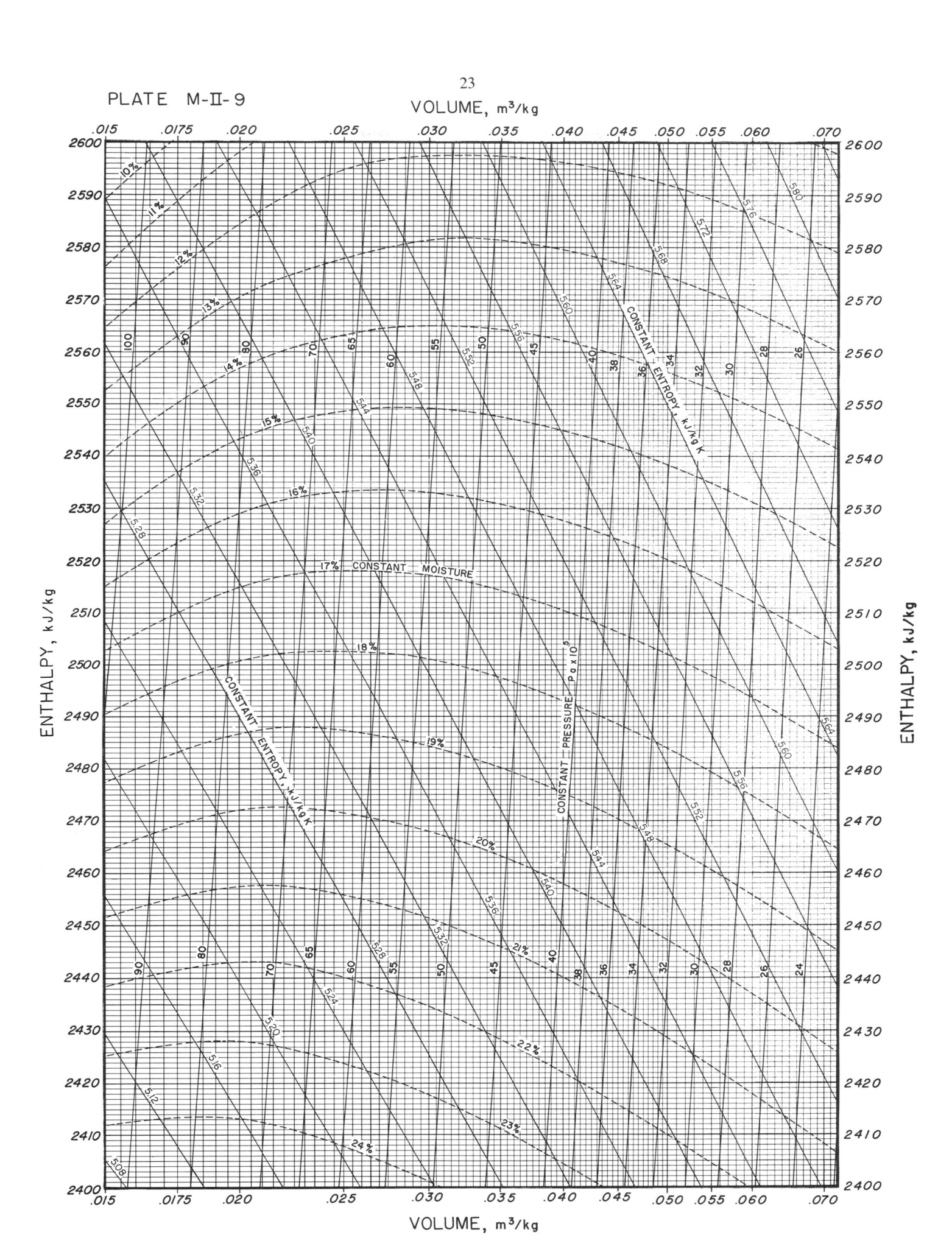

PLATE M-II-10

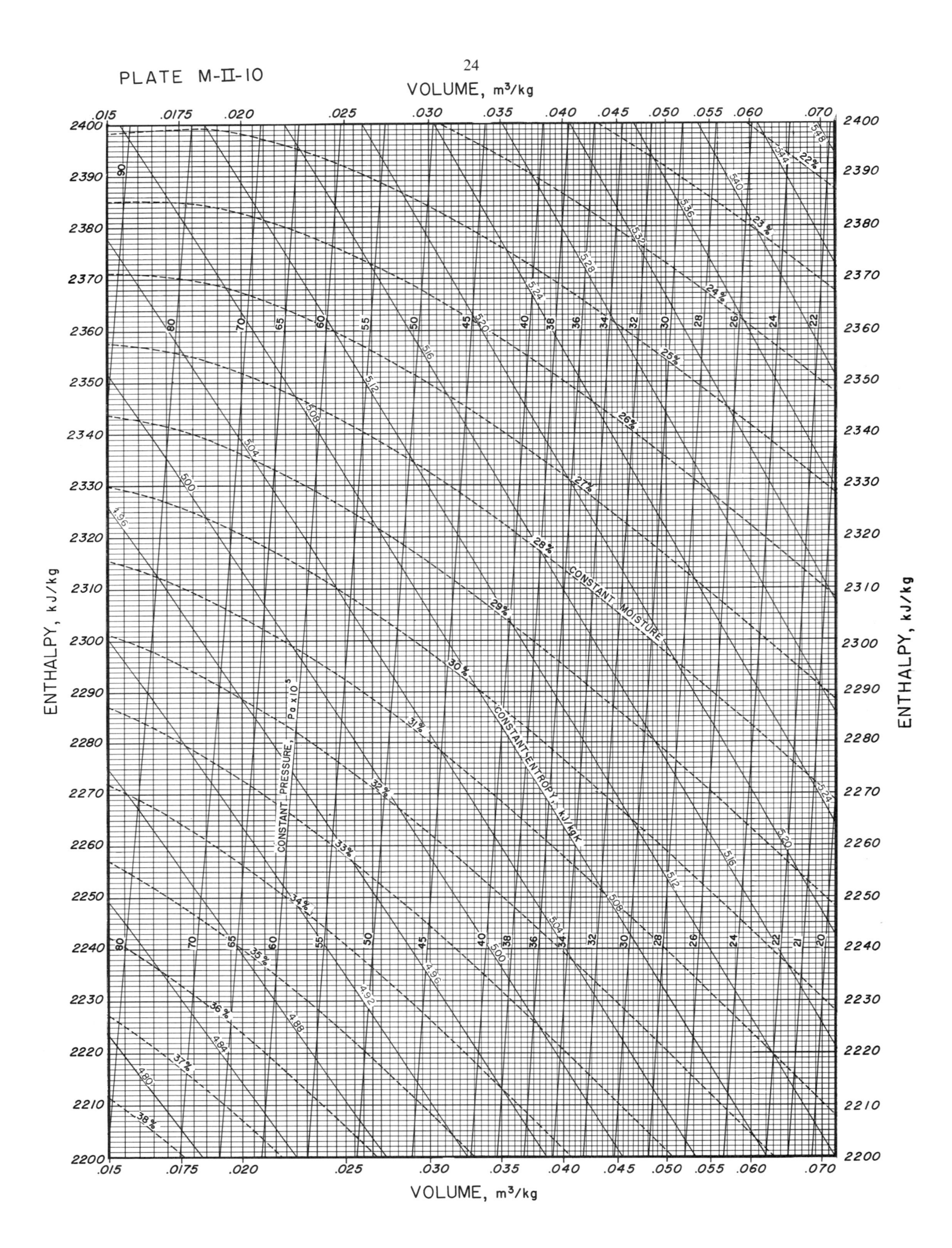

PLATE M-II-II

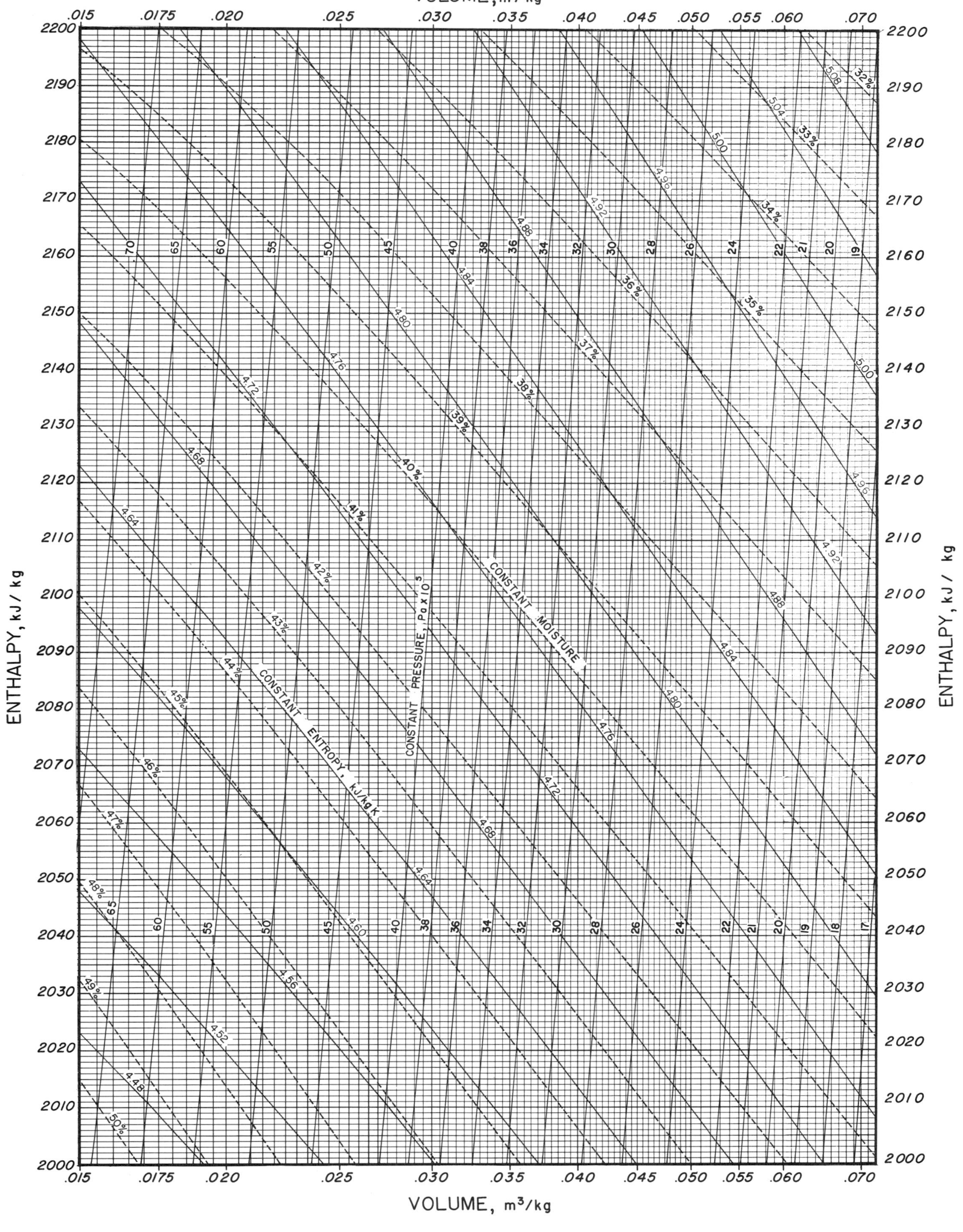
VOLUME, m³/kg
.015 .0175 .020 .025 .030 .035 .040 .045 .050 .055 .060 .070
ENTHALPY, kJ / kg
2000 2010 2020 2030 2040 2050 2060 2070 2080 2090 2100 2110 2120 2130 2140 2150 2160 2170 2180 2190 2200
CONSTANT PRESSURE, Pa x 10⁵
70 65 60 55 50 45 40 38 36 34 32 30 28 26 24 22 21 20 19 18 17
CONSTANT MOISTURE
32% 33% 34% 35% 36% 37% 38% 39% 40% 41% 42% 43% 44% 45% 46% 47% 48% 49% 50%
CONSTANT ENTROPY, kJ/kg K
4.48 4.52 4.56 4.60 4.64 4.68 4.72 4.76 4.80 4.84 4.88 4.92 4.96 5.00 5.04 5.08

PLATE M-II-12

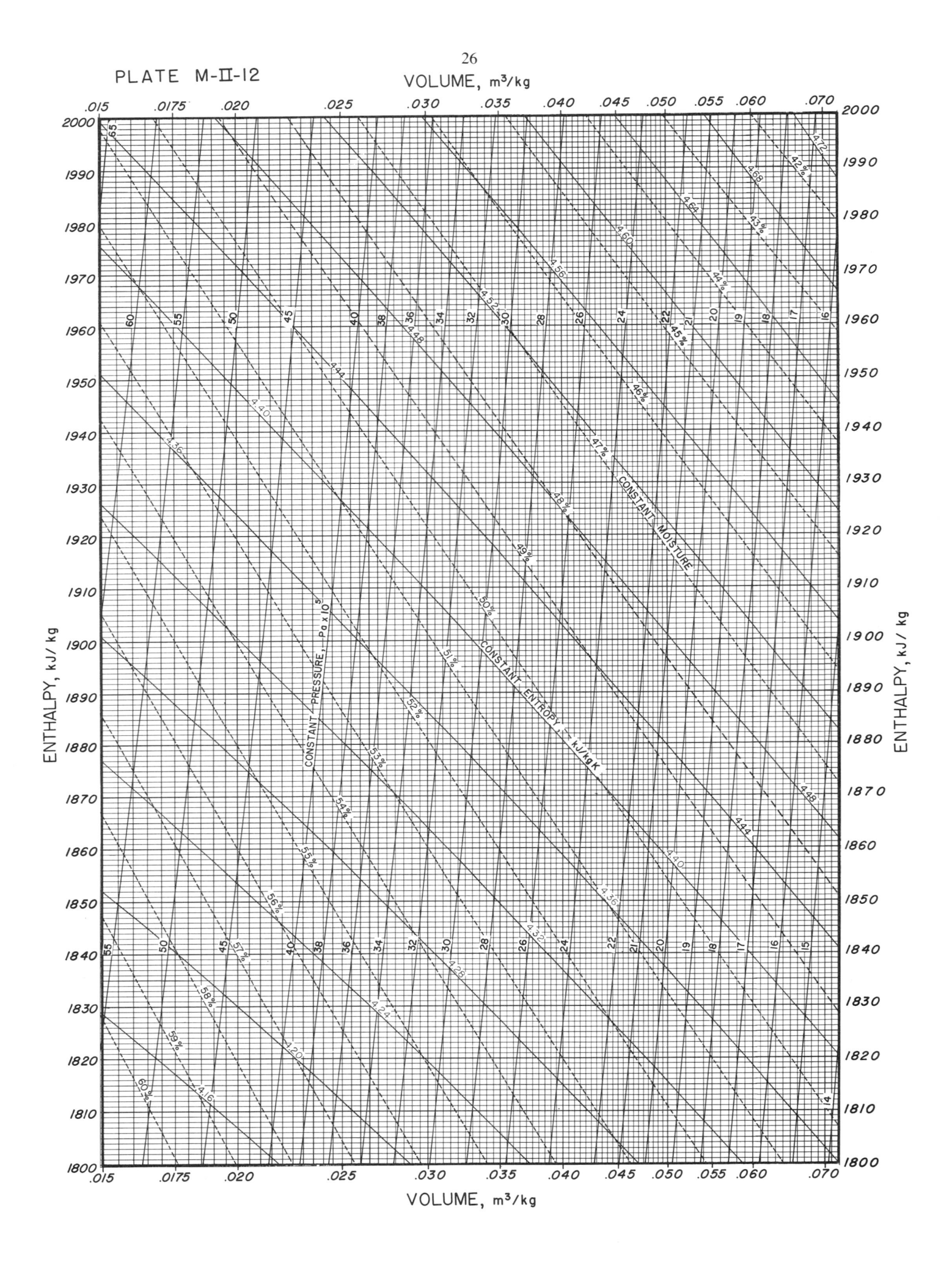

PLATE M-III-1

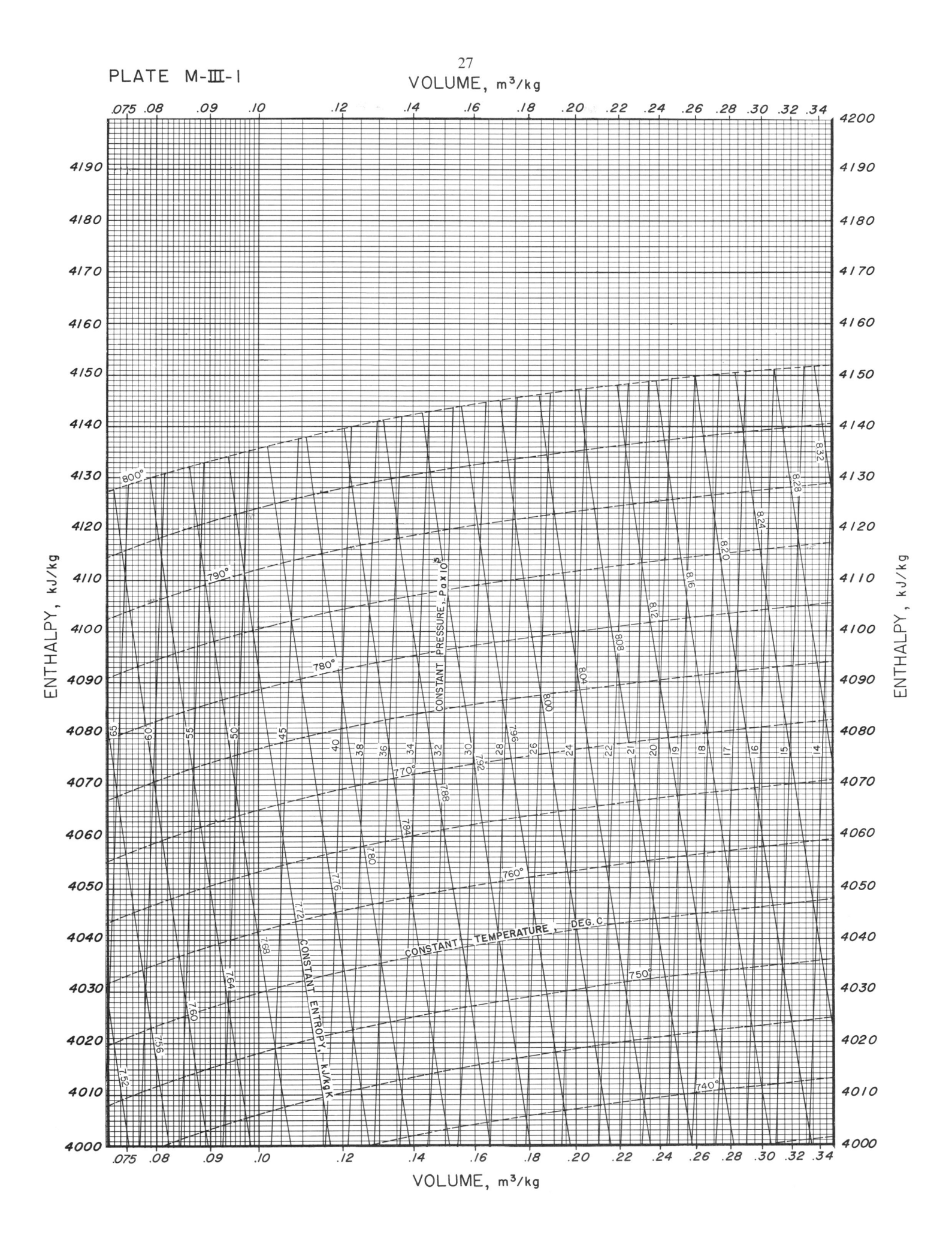

PLATE M-III-2

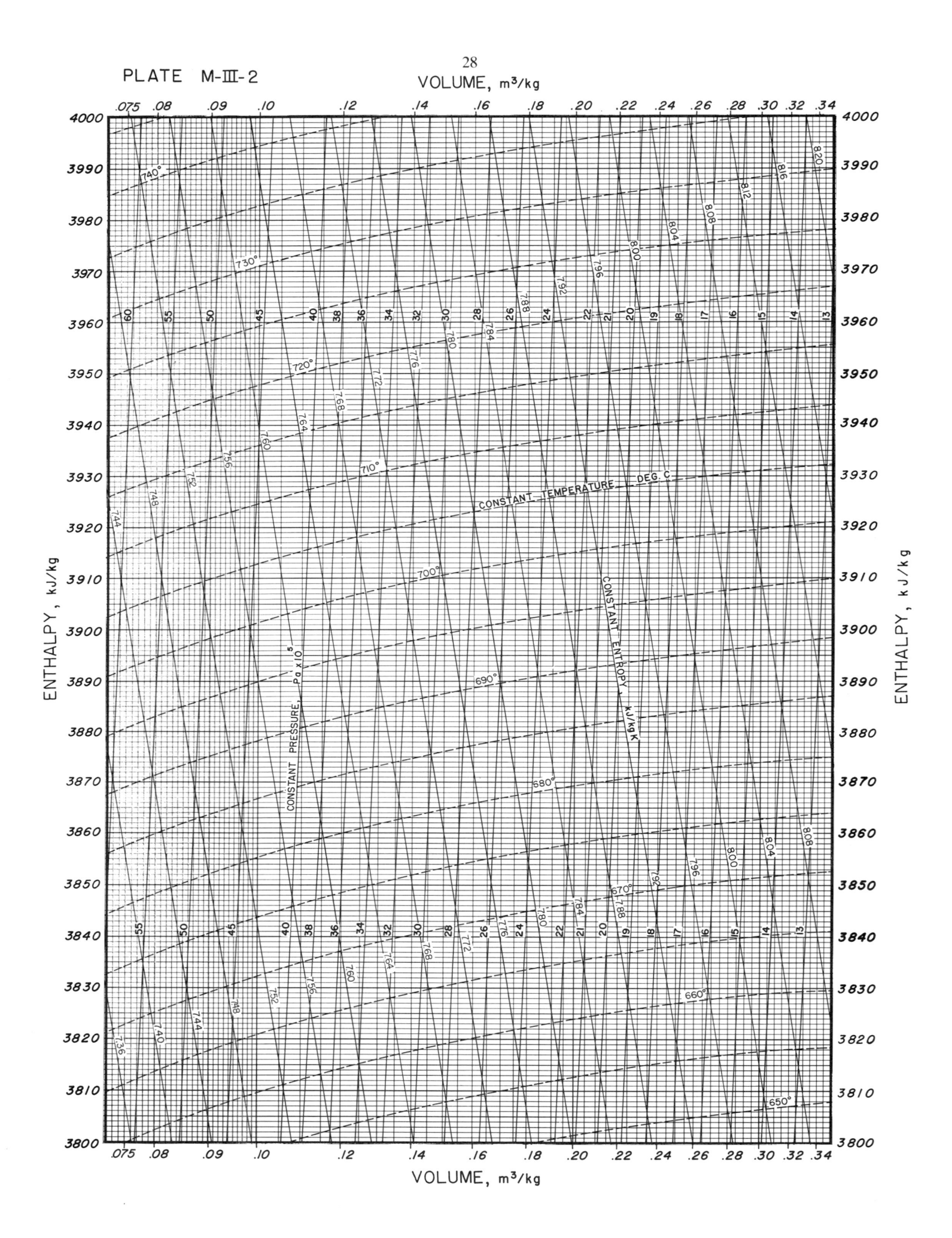

PLATE M-III-3

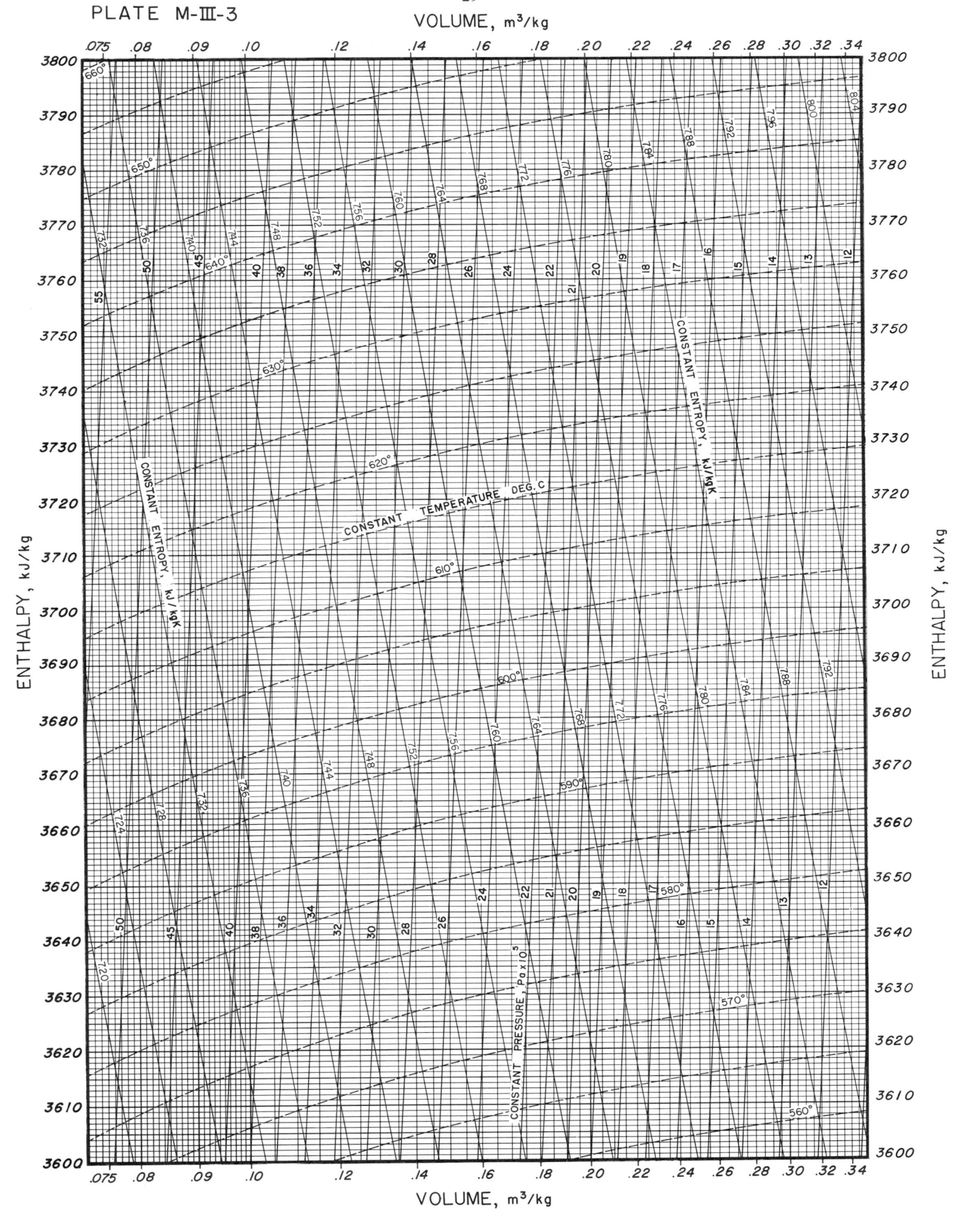

PLATE M-III-4

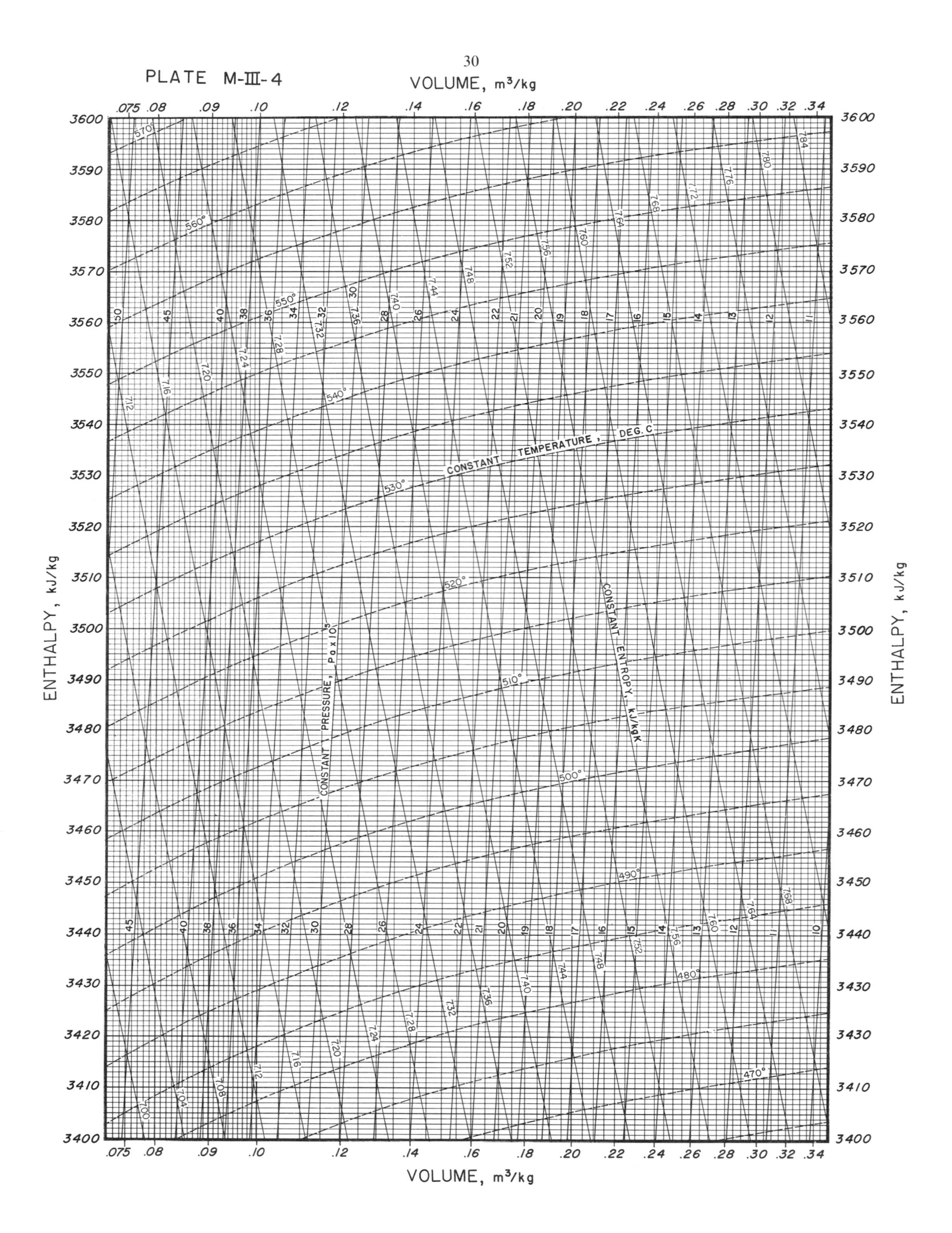

PLATE M-III-5

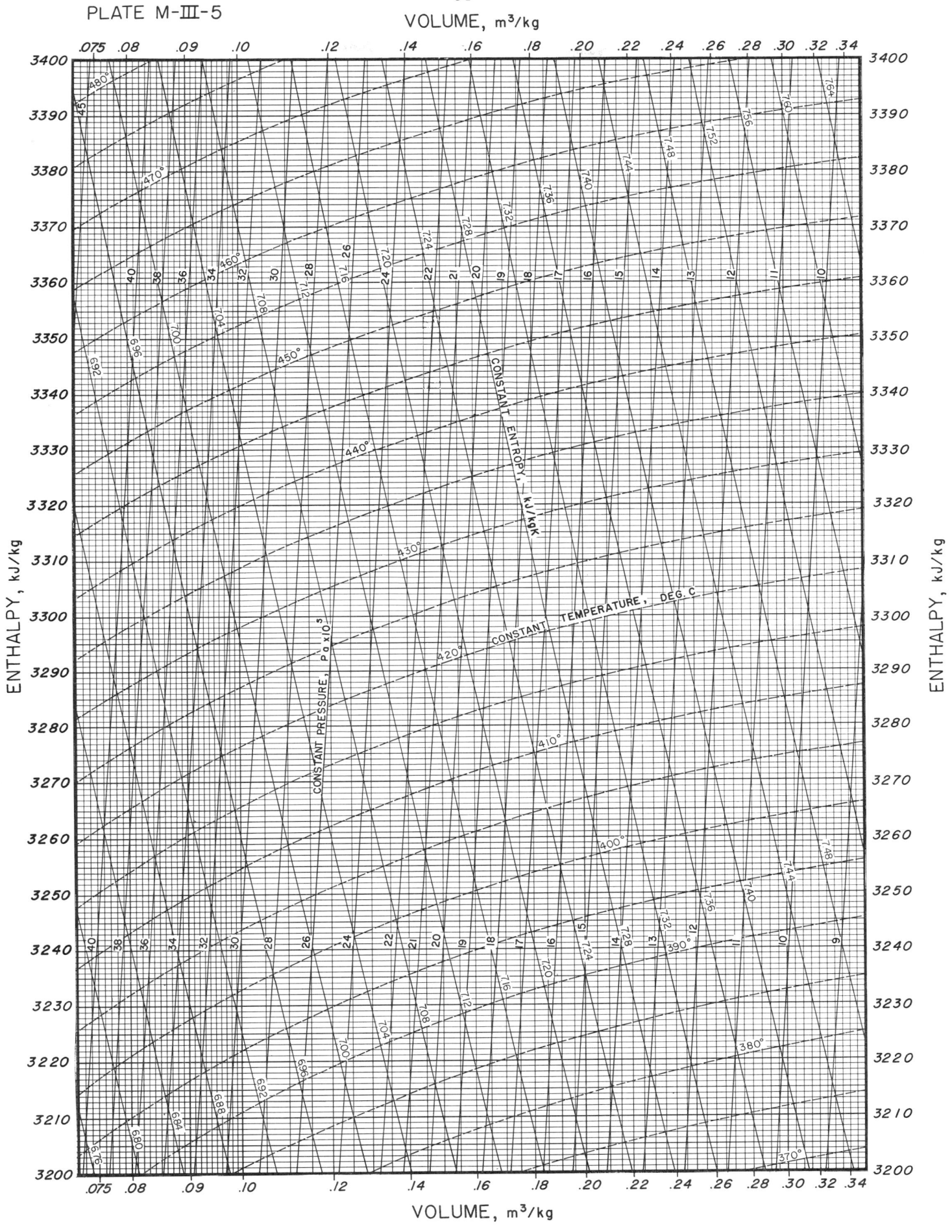

PLATE M-III-6

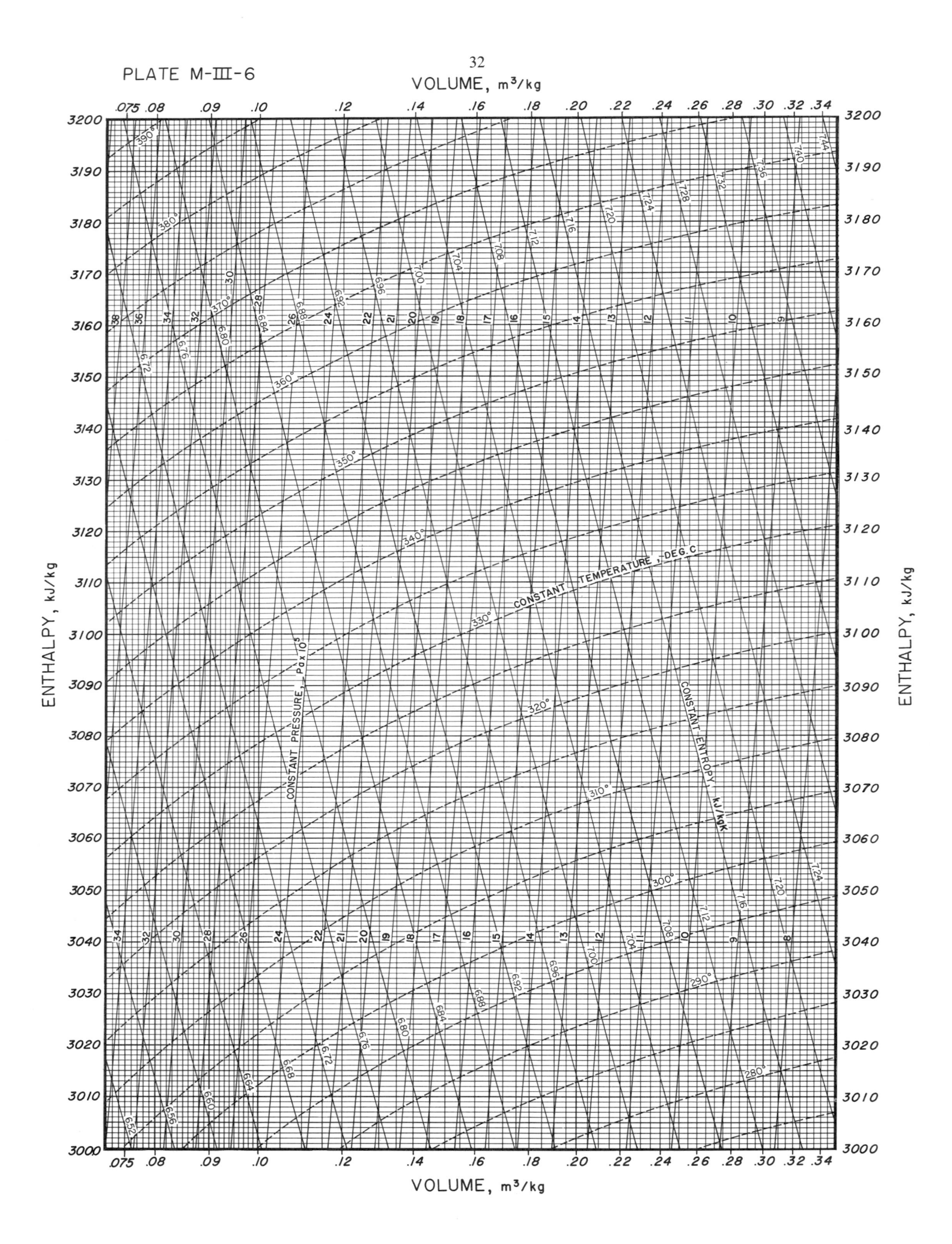

PLATE M-III-7

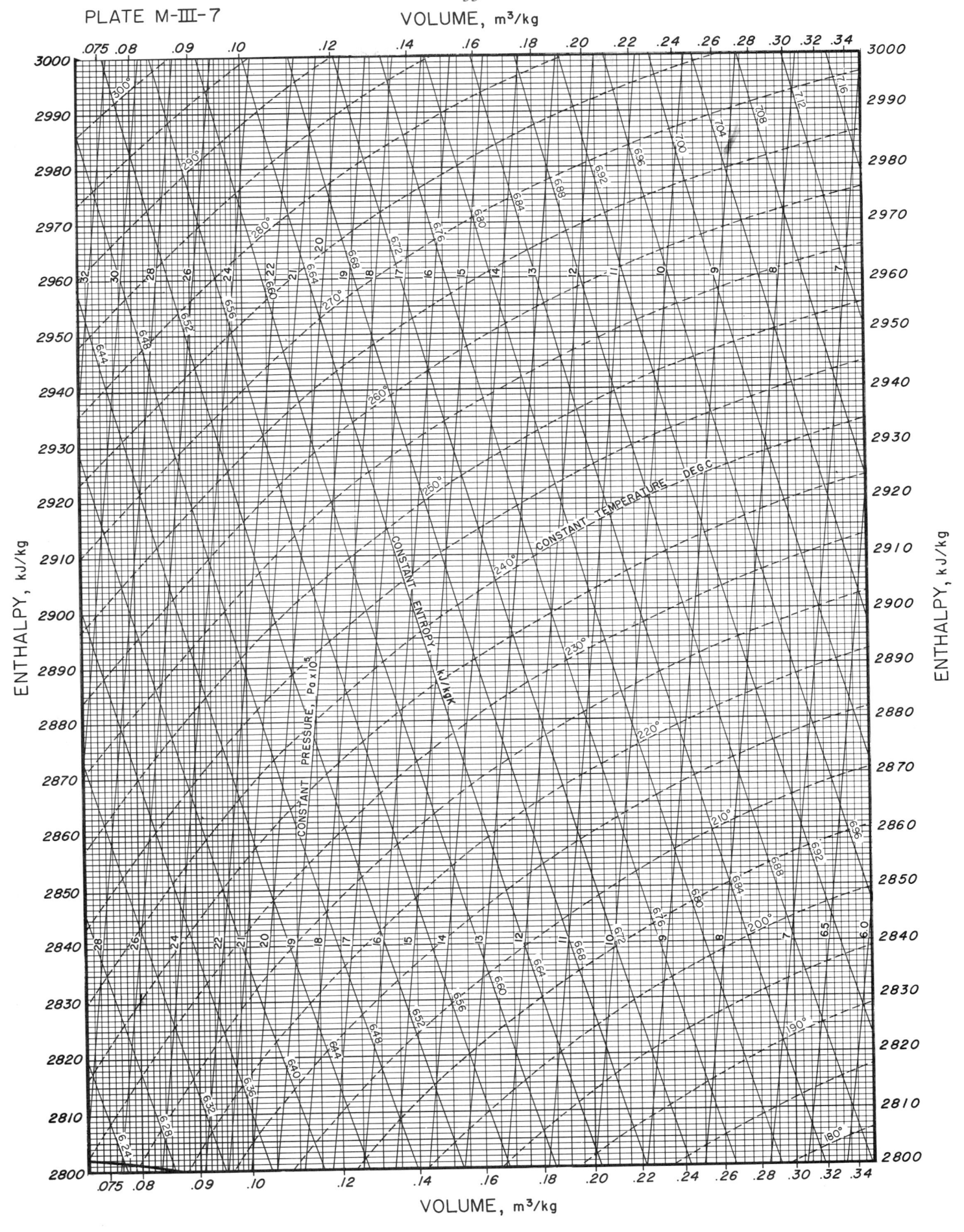

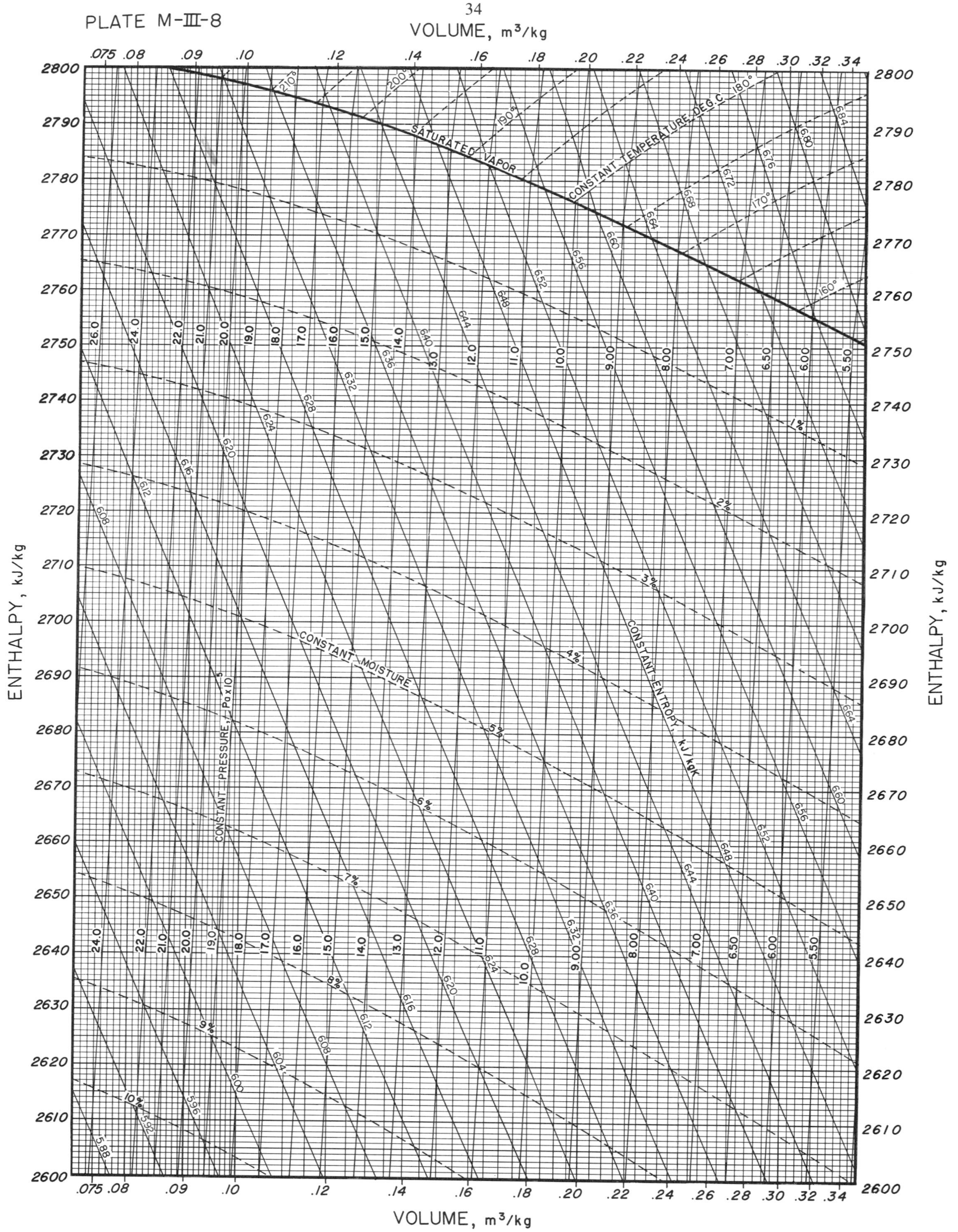
PLATE M-III-8
VOLUME, m³/kg
ENTHALPY, kJ/kg
SATURATED VAPOR
CONSTANT TEMPERATURE, DEG. C.
CONSTANT MOISTURE
CONSTANT PRESSURE, Pa x 10^5
CONSTANT ENTROPY, kJ/kgK

PLATE M-III-9

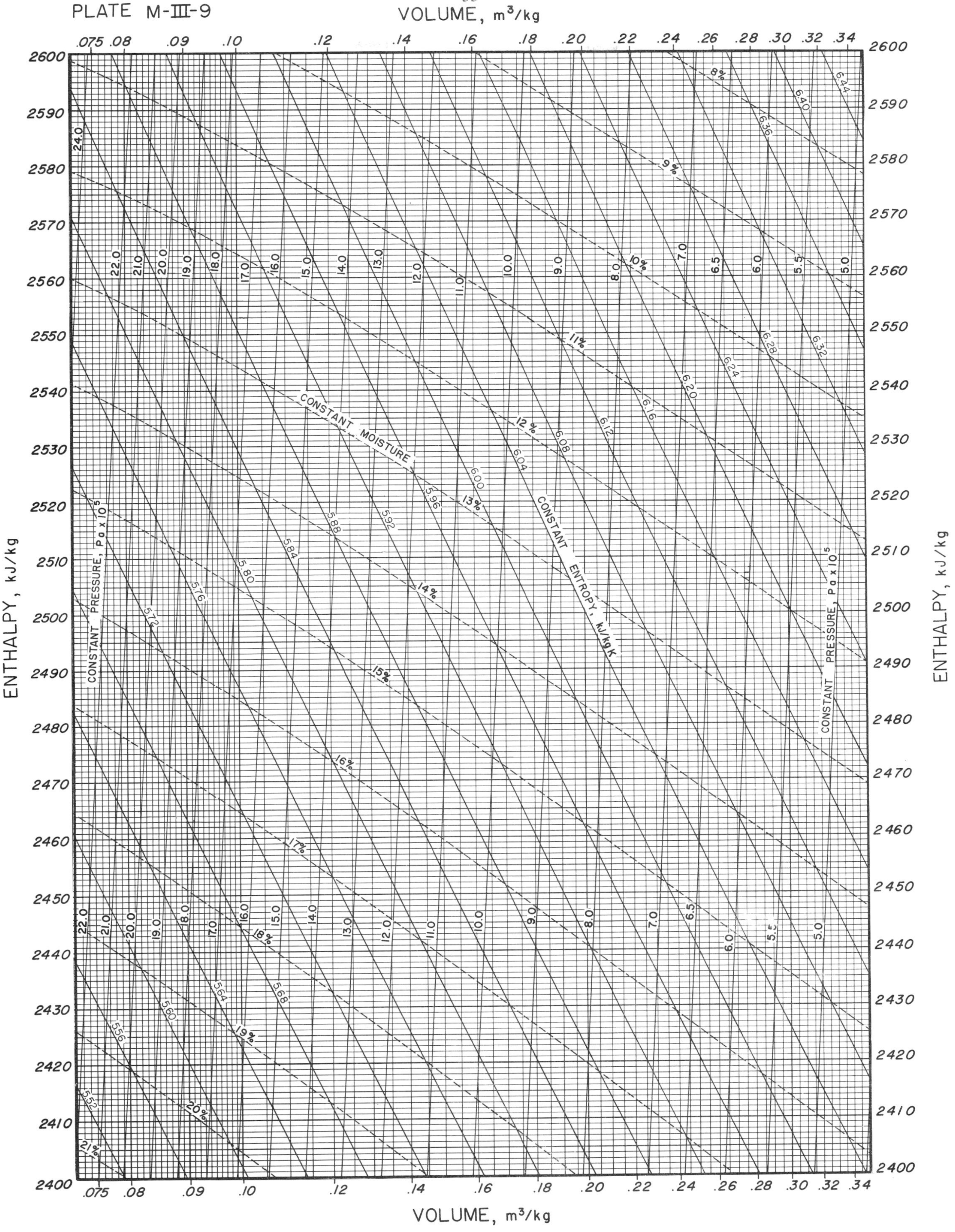
VOLUME, m³/kg
.075 .08 .09 .10 .12 .14 .16 .18 .20 .22 .24 .26 .28 .30 .32 .34
ENTHALPY, kJ/kg
2400 2410 2420 2430 2440 2450 2460 2470 2480 2490 2500 2510 2520 2530 2540 2550 2560 2570 2580 2590 2600
CONSTANT PRESSURE, Pa x 10⁵
24.0 22.0 21.0 20.0 19.0 18.0 17.0 16.0 15.0 14.0 13.0 12.0 11.0 10.0 9.0 8.0 7.0 6.5 6.0 5.5 5.0
CONSTANT MOISTURE
8% 9% 10% 11% 12% 13% 14% 15% 16% 17% 18% 19% 20% 21%
CONSTANT ENTROPY, kJ/kg K
5.52 5.56 5.60 5.64 5.68 5.72 5.76 5.80 5.84 5.88 5.92 5.96 6.00 6.04 6.08 6.12 6.16 6.20 6.24 6.28 6.32 6.36 6.40 6.44

PLATE M-III-10

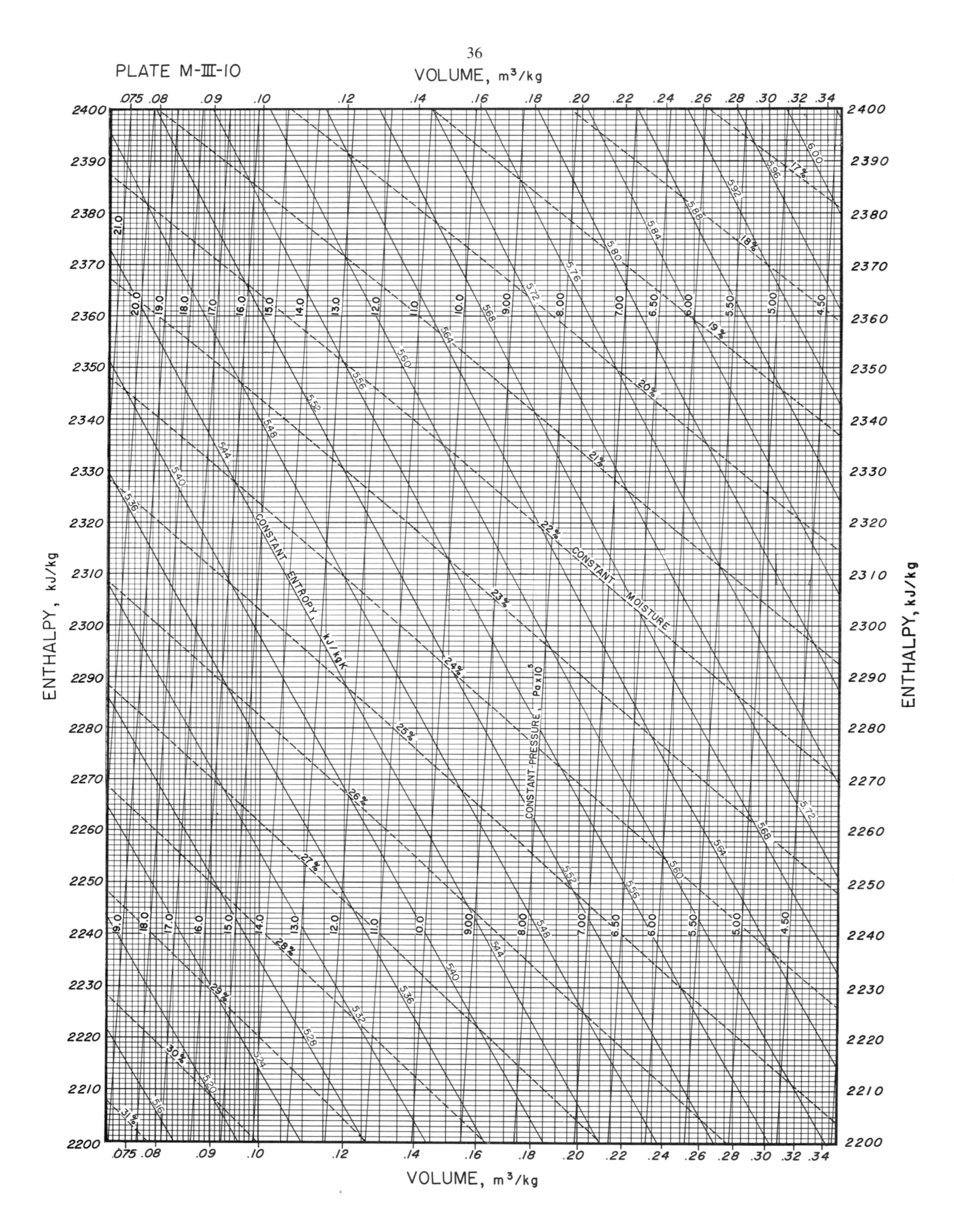

PLATE M-III-11

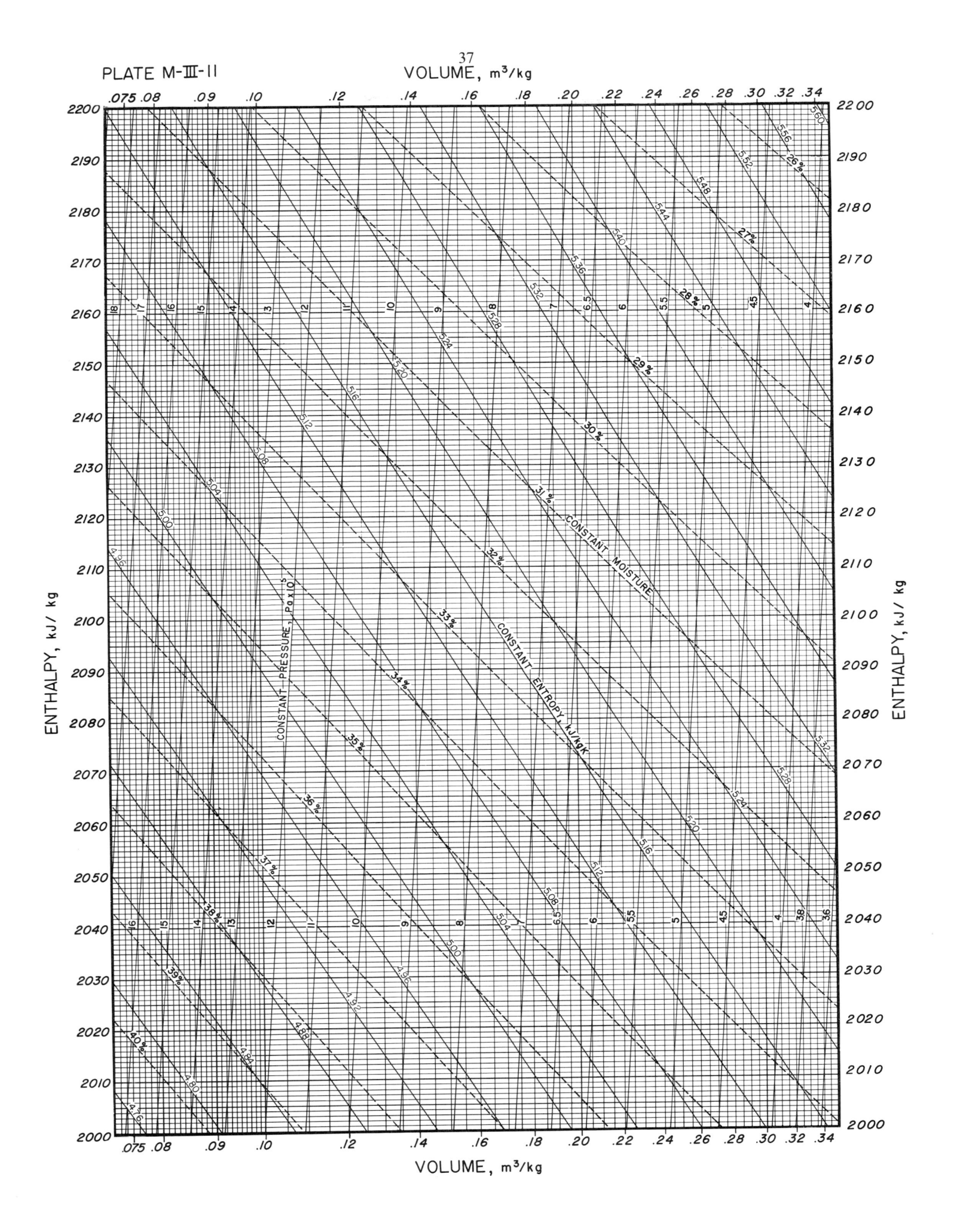

PLATE M-III-12

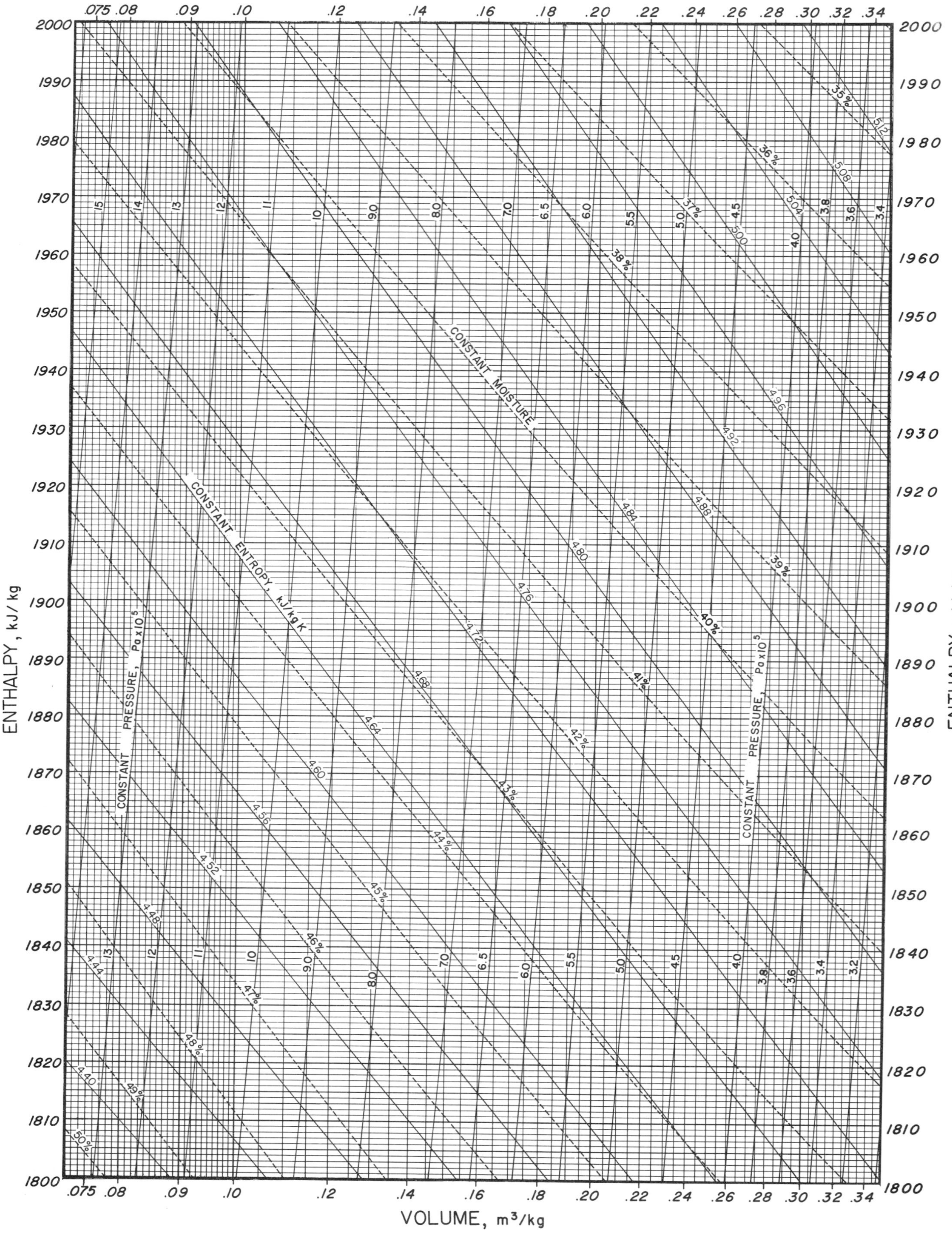

PLATE M-IV-I

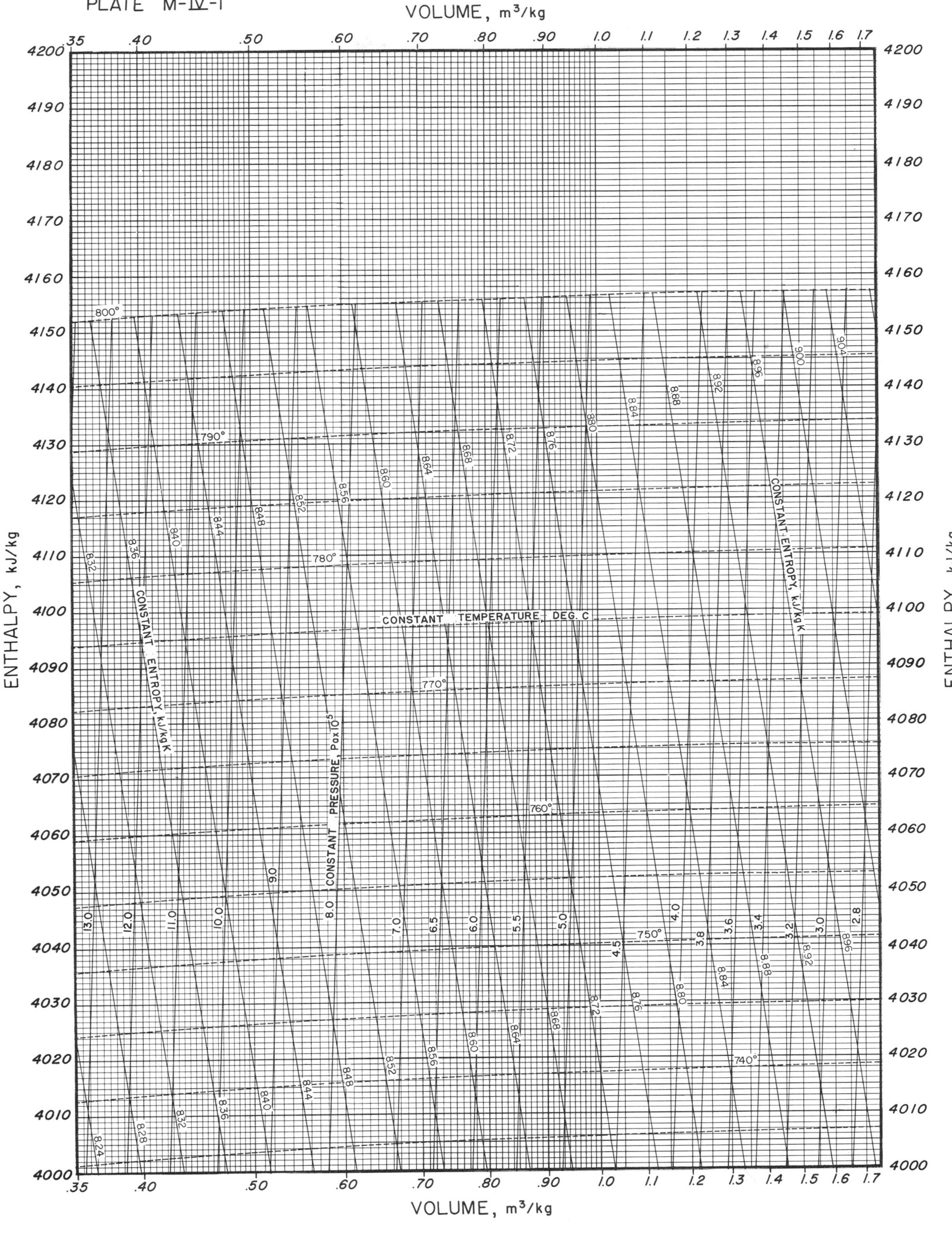
VOLUME, m³/kg
.35
.40
.50
.60
.70
.80
.90
1.0
1.1
1.2
1.3
1.4
1.5
1.6
1.7
ENTHALPY, kJ/kg
4200
4190
4180
4170
4160
4150
4140
4130
4120
4110
4100
4090
4080
4070
4060
4050
4040
4030
4020
4010
4000
800°
790°
780°
770°
760°
750°
740°
CONSTANT TEMPERATURE, DEG. C
CONSTANT ENTROPY, kJ/kg K
CONSTANT PRESSURE, Pa x 10⁵
13.0
12.0
11.0
10.0
9.0
8.0
7.0
6.5
6.0
5.5
5.0
4.5
4.0
3.8
3.6
3.4
3.2
3.0
2.8
8.24
8.28
8.32
8.36
8.40
8.44
8.48
8.52
8.56
8.60
8.64
8.68
8.72
8.76
8.80
8.84
8.88
8.92
8.96
9.00
9.04

PLATE M-IV-2

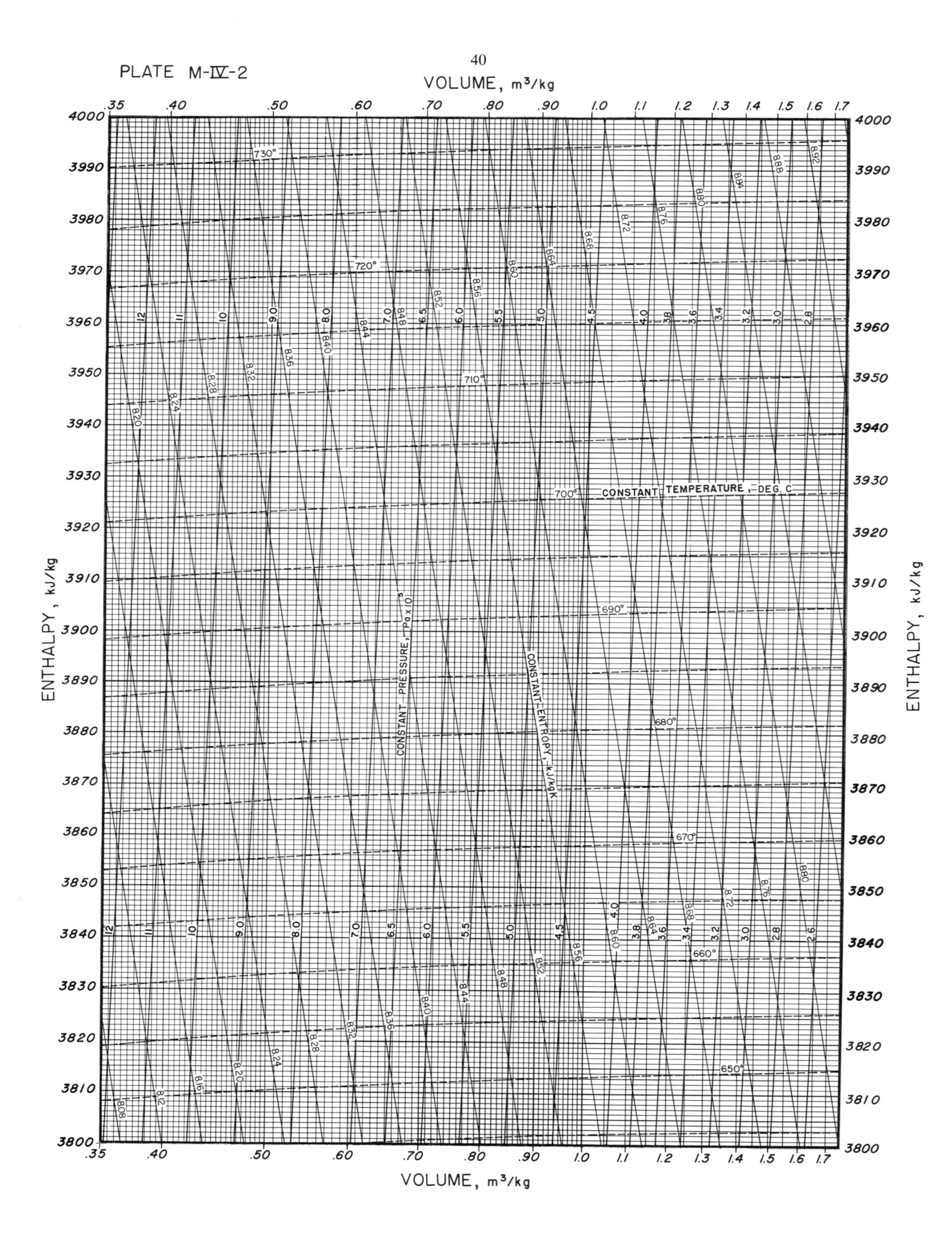

PLATE M-IV-3

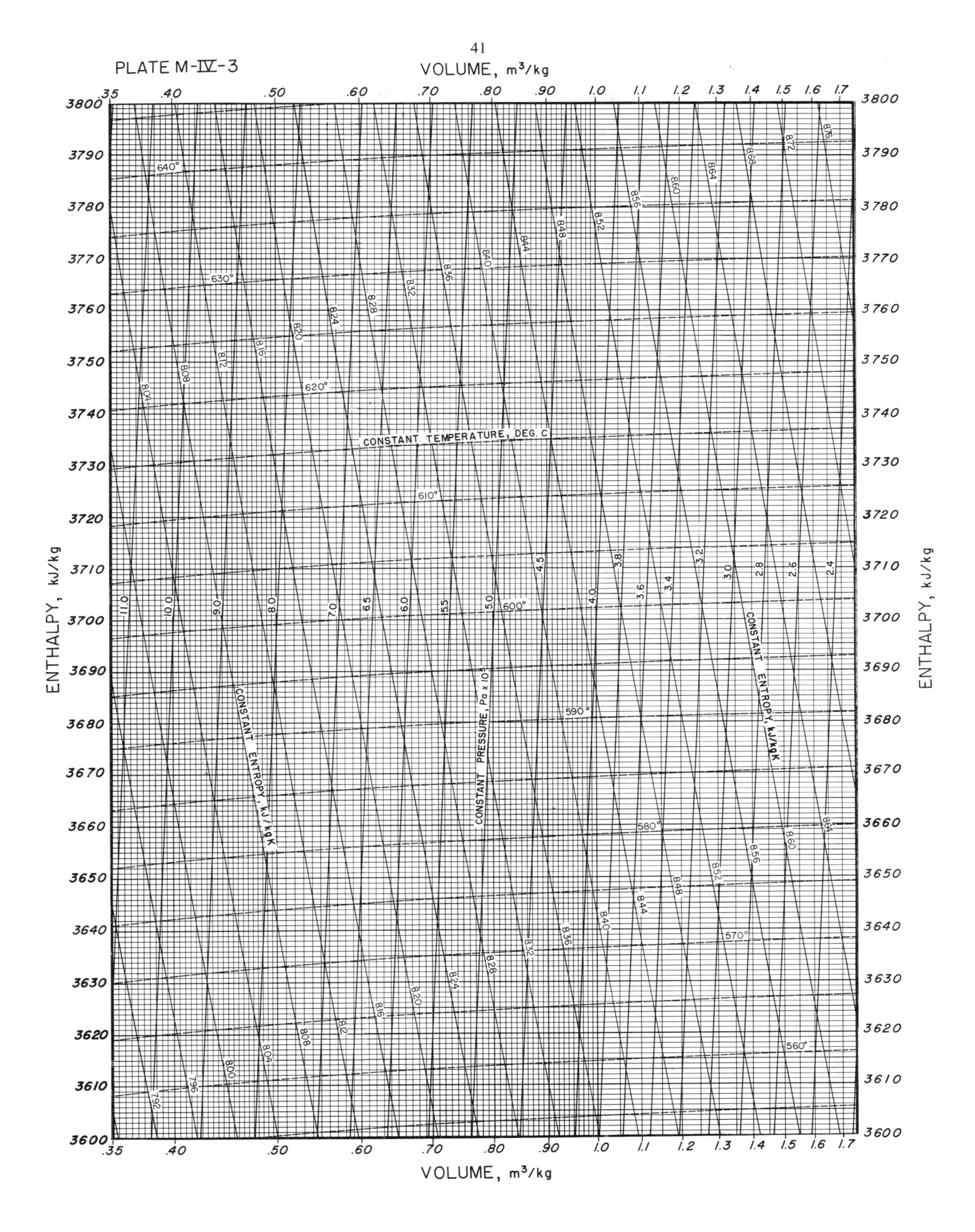

PLATE M-IV-4

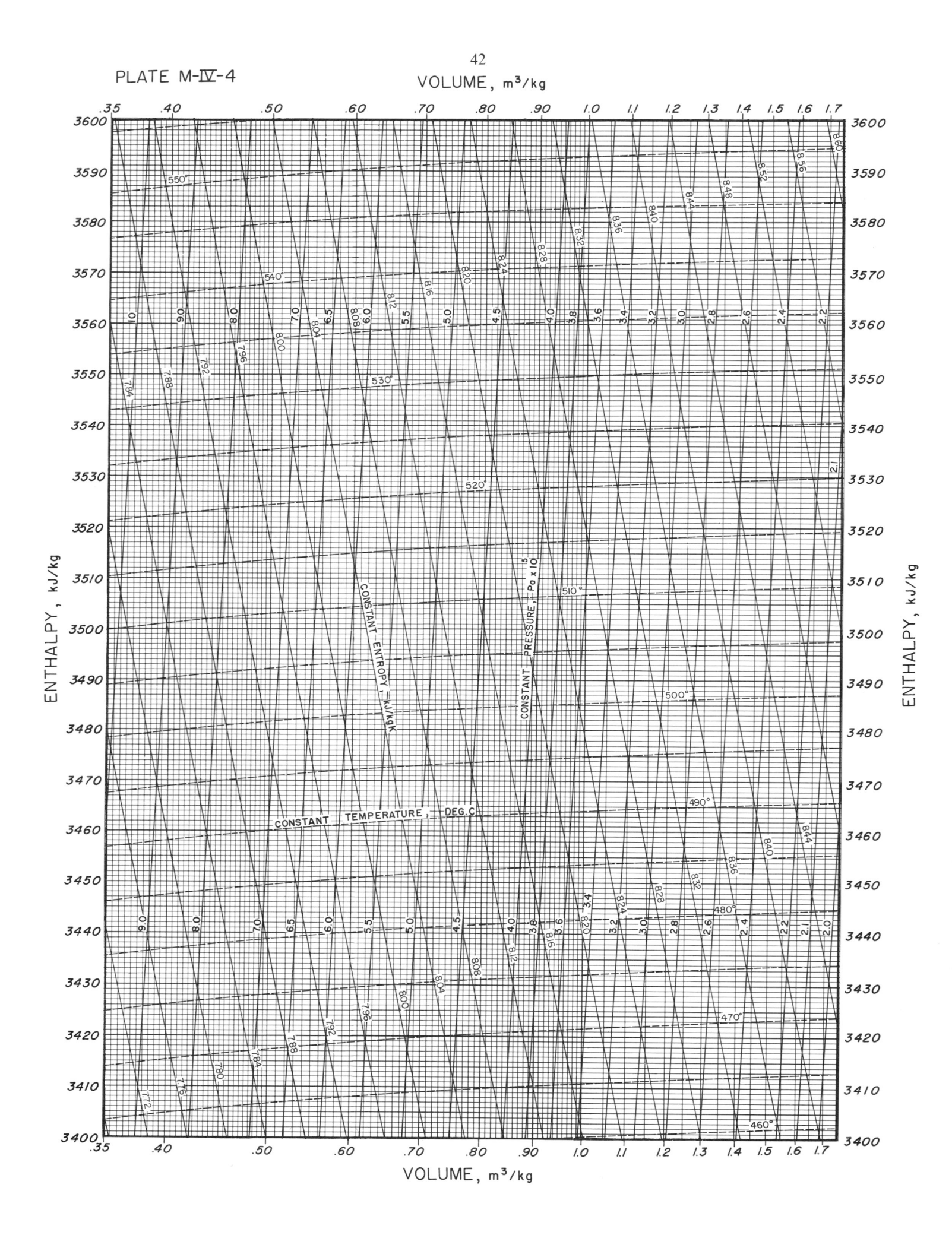

PLATE M-IV-5

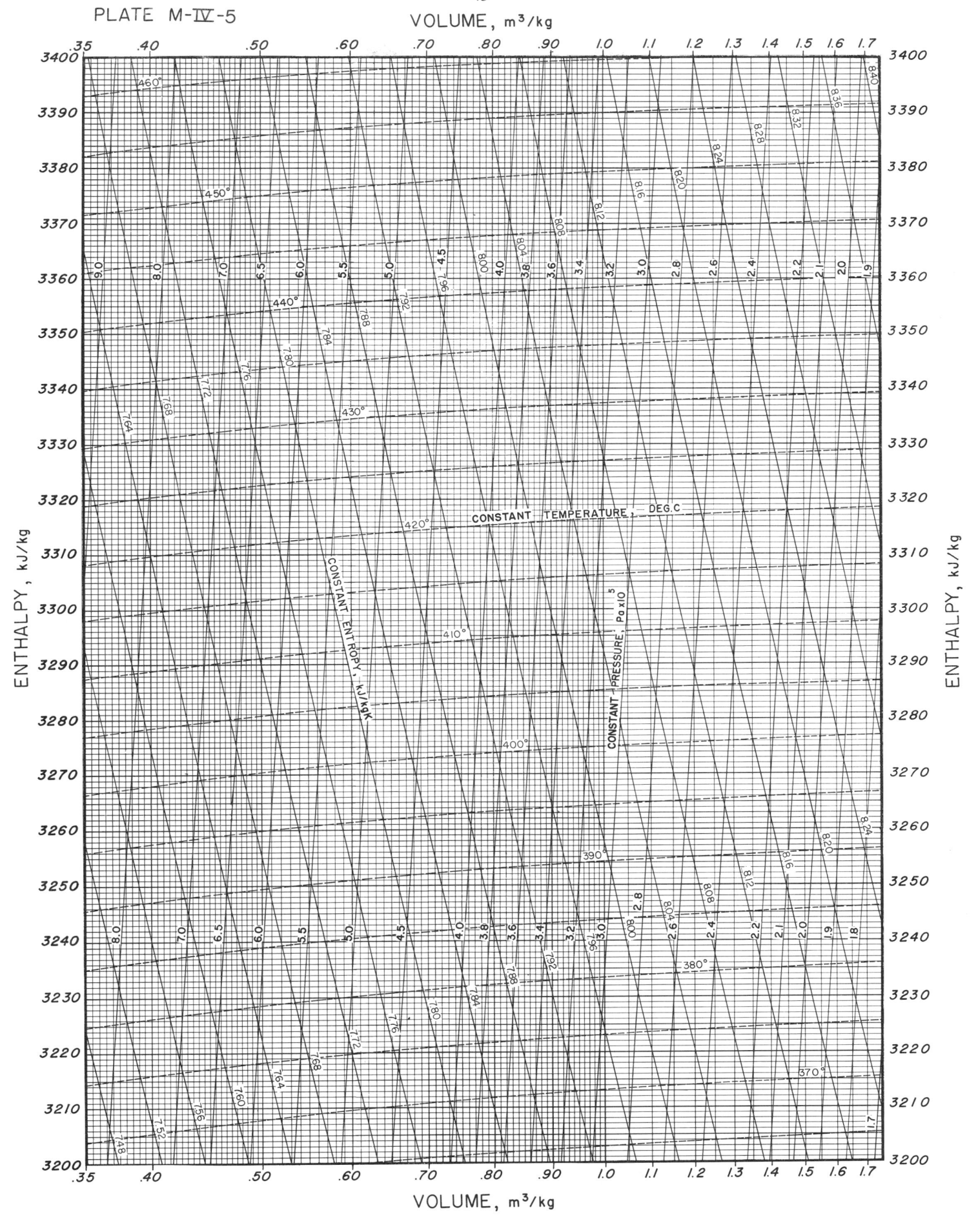

PLATE M-IV-6

PLATE M-IV-7

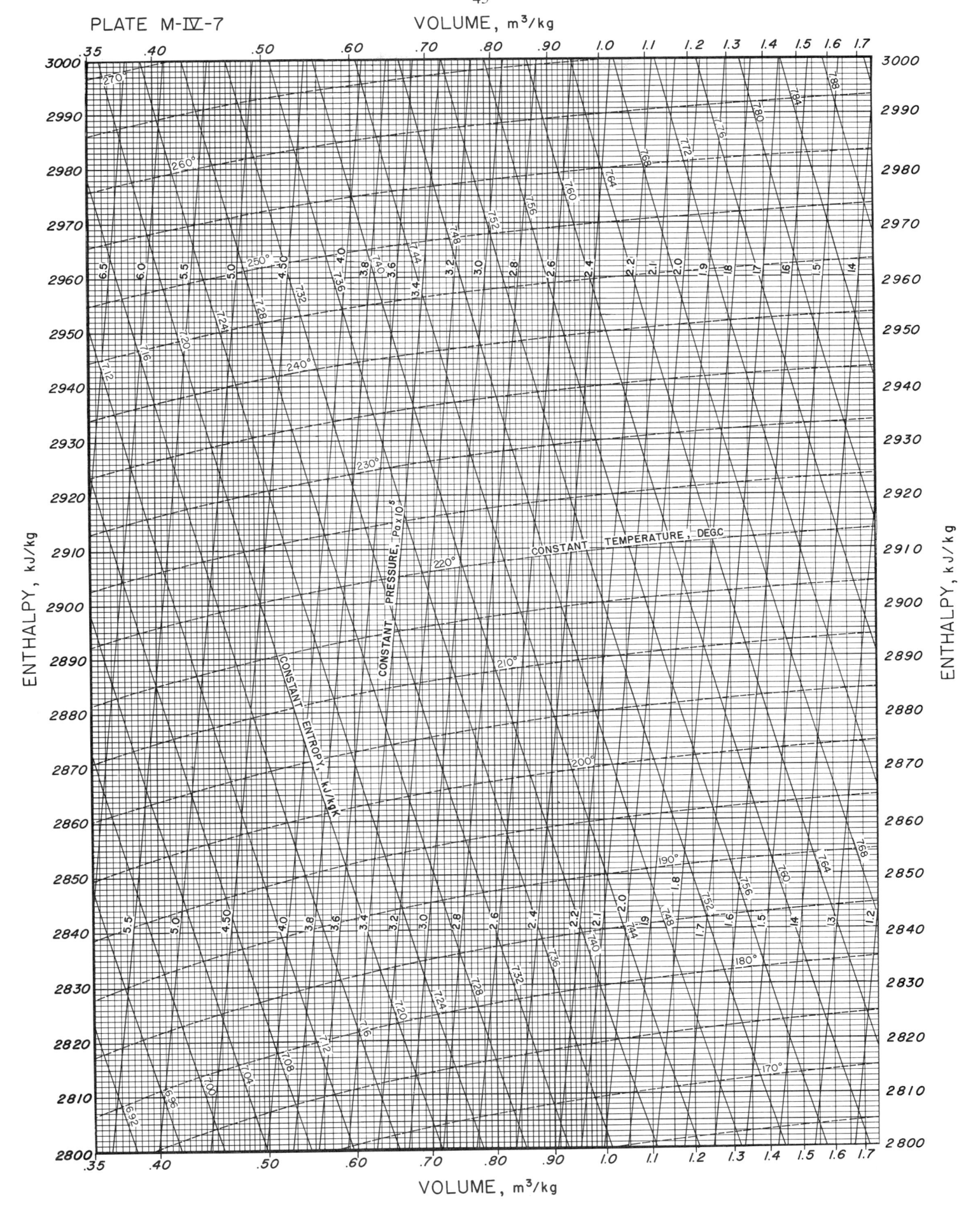

PLATE M-IV-8

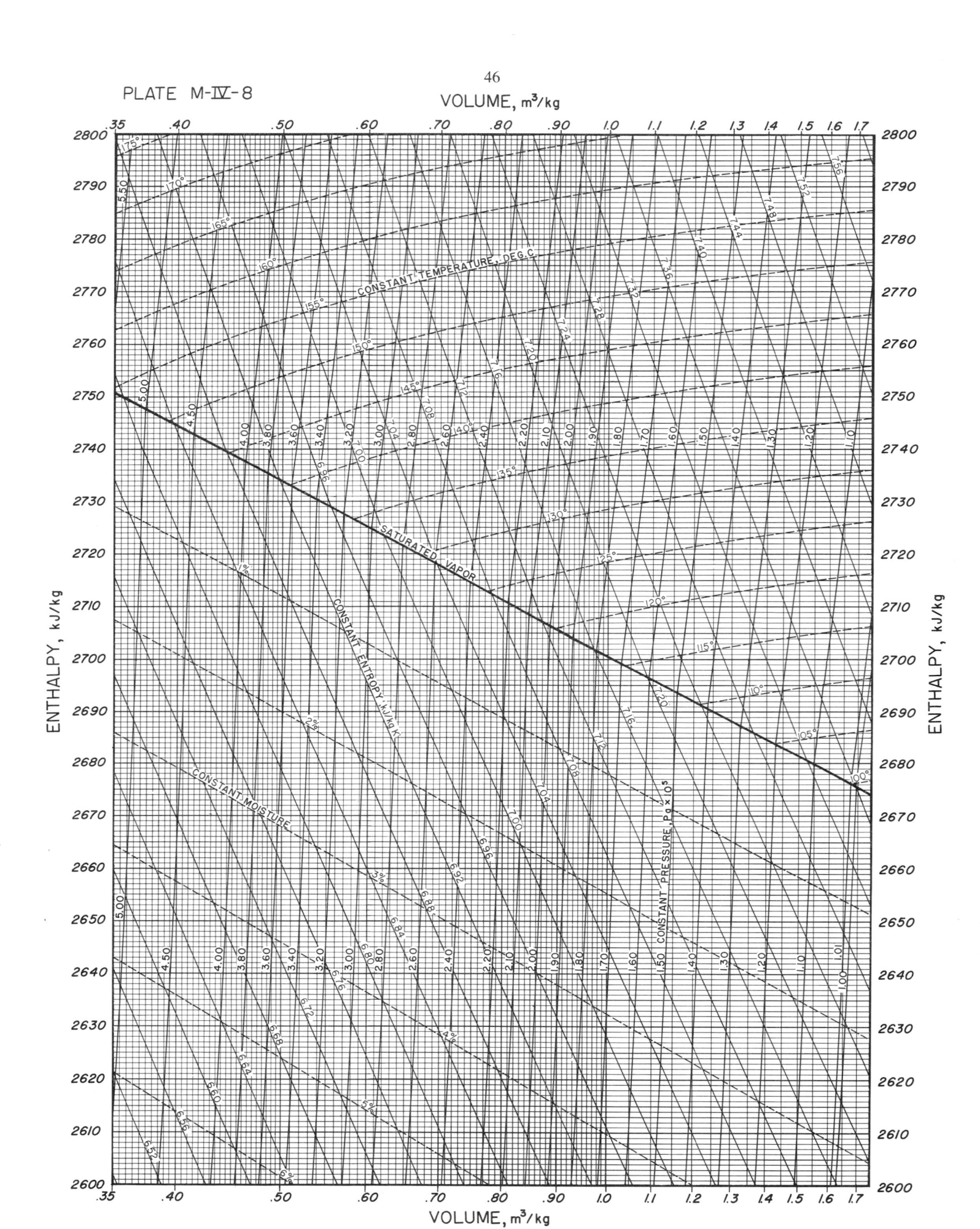

PLATE M-IV-9

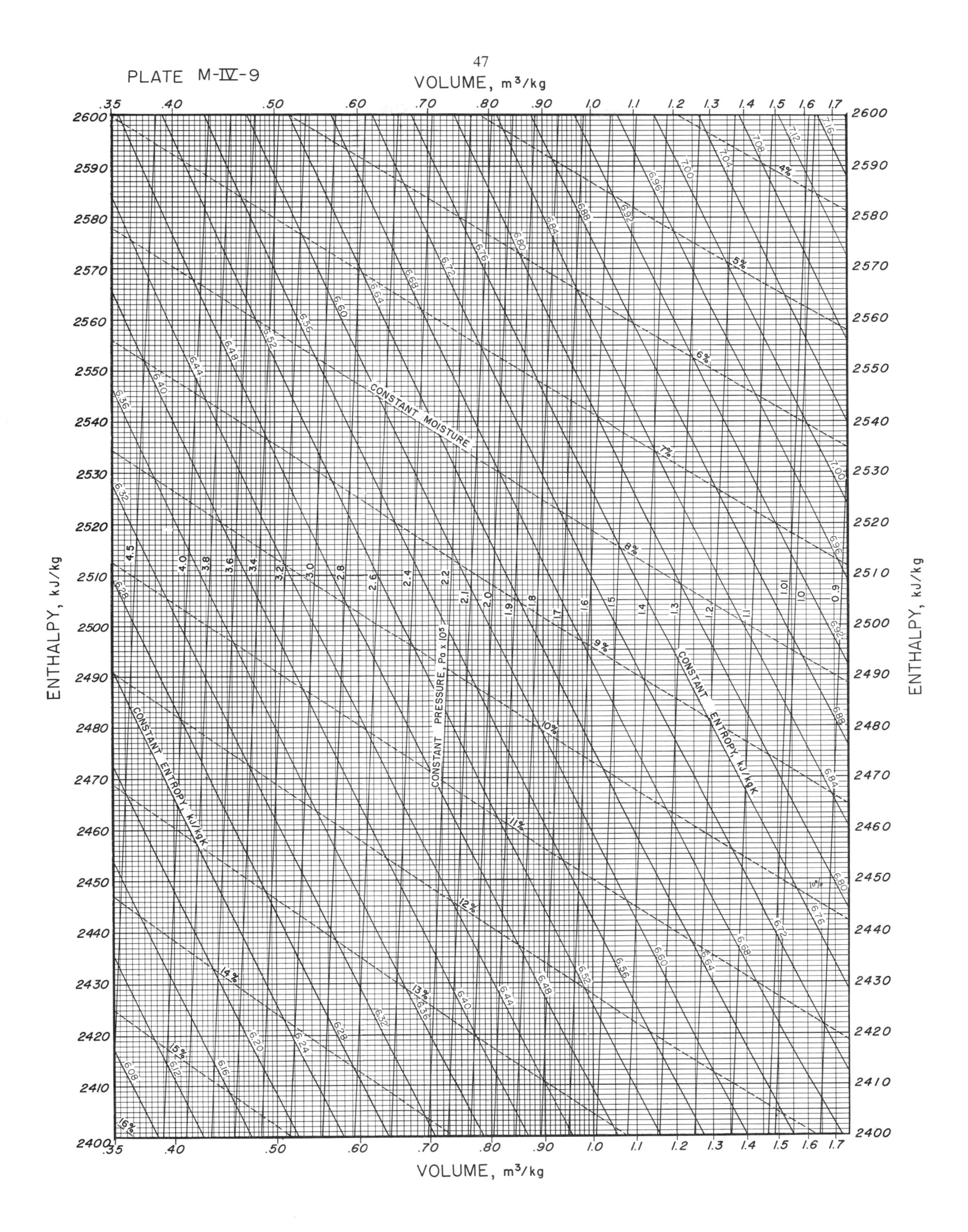

PLATE M-IV-10

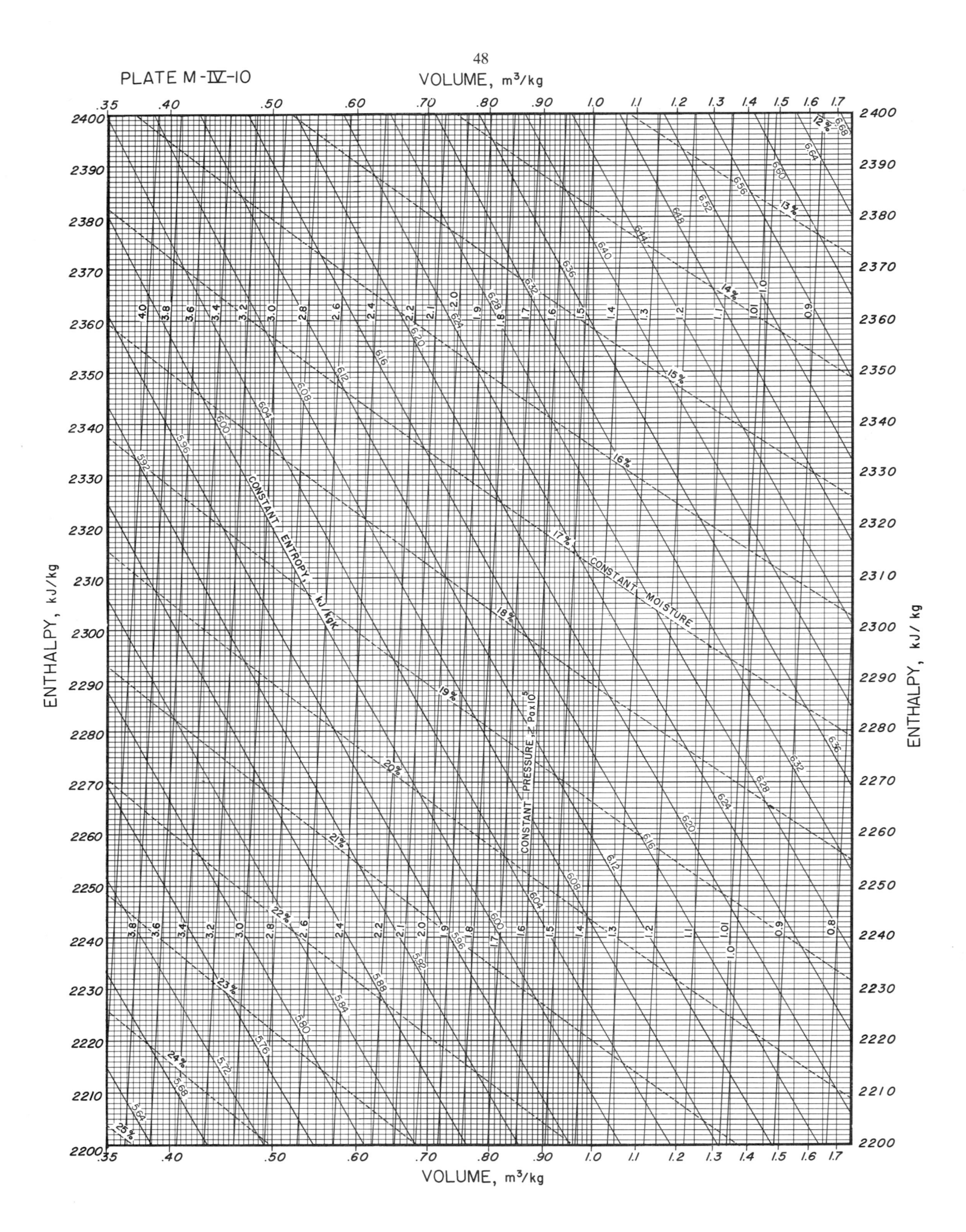

PLATE M-IV-11

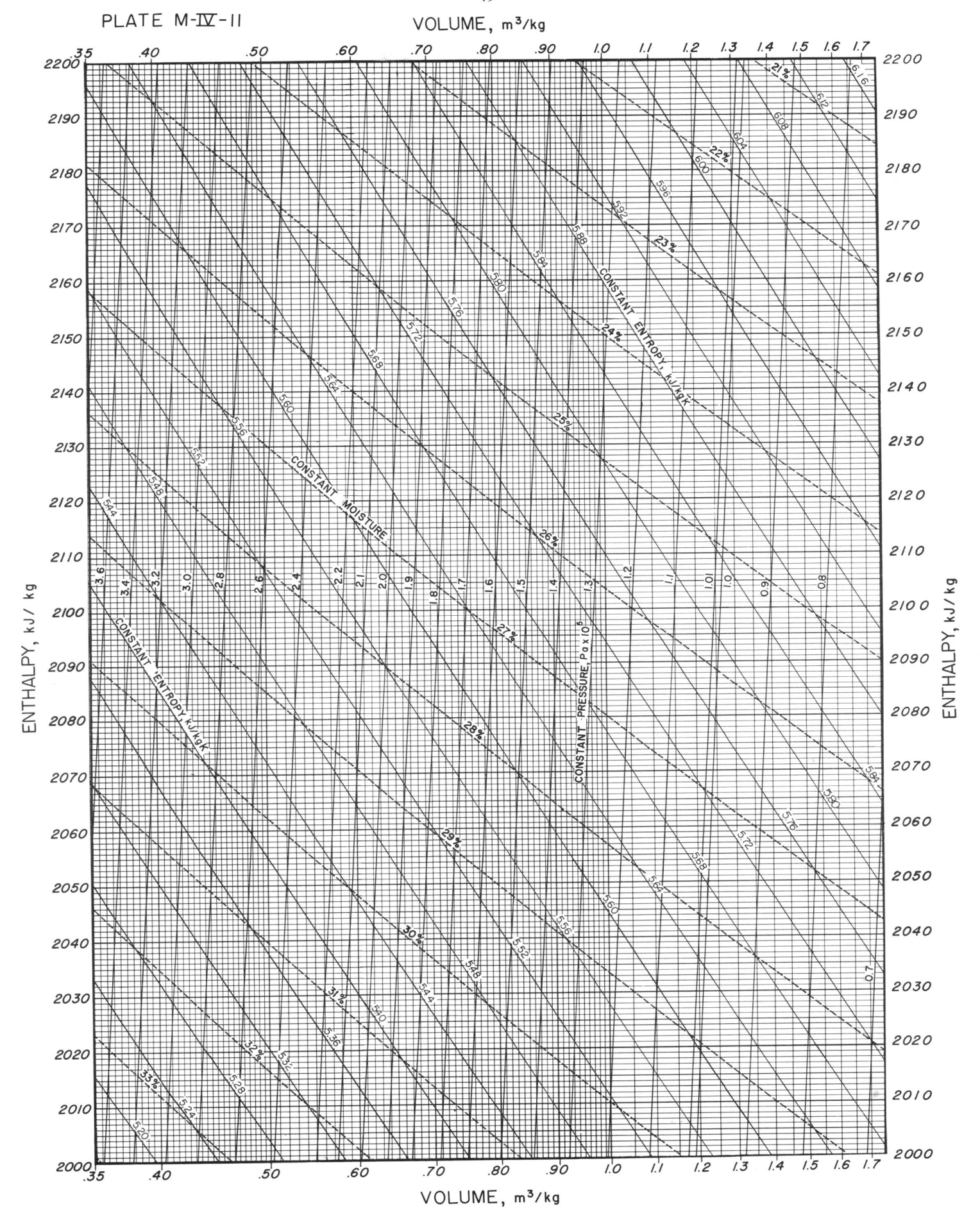

PLATE M-IV-12

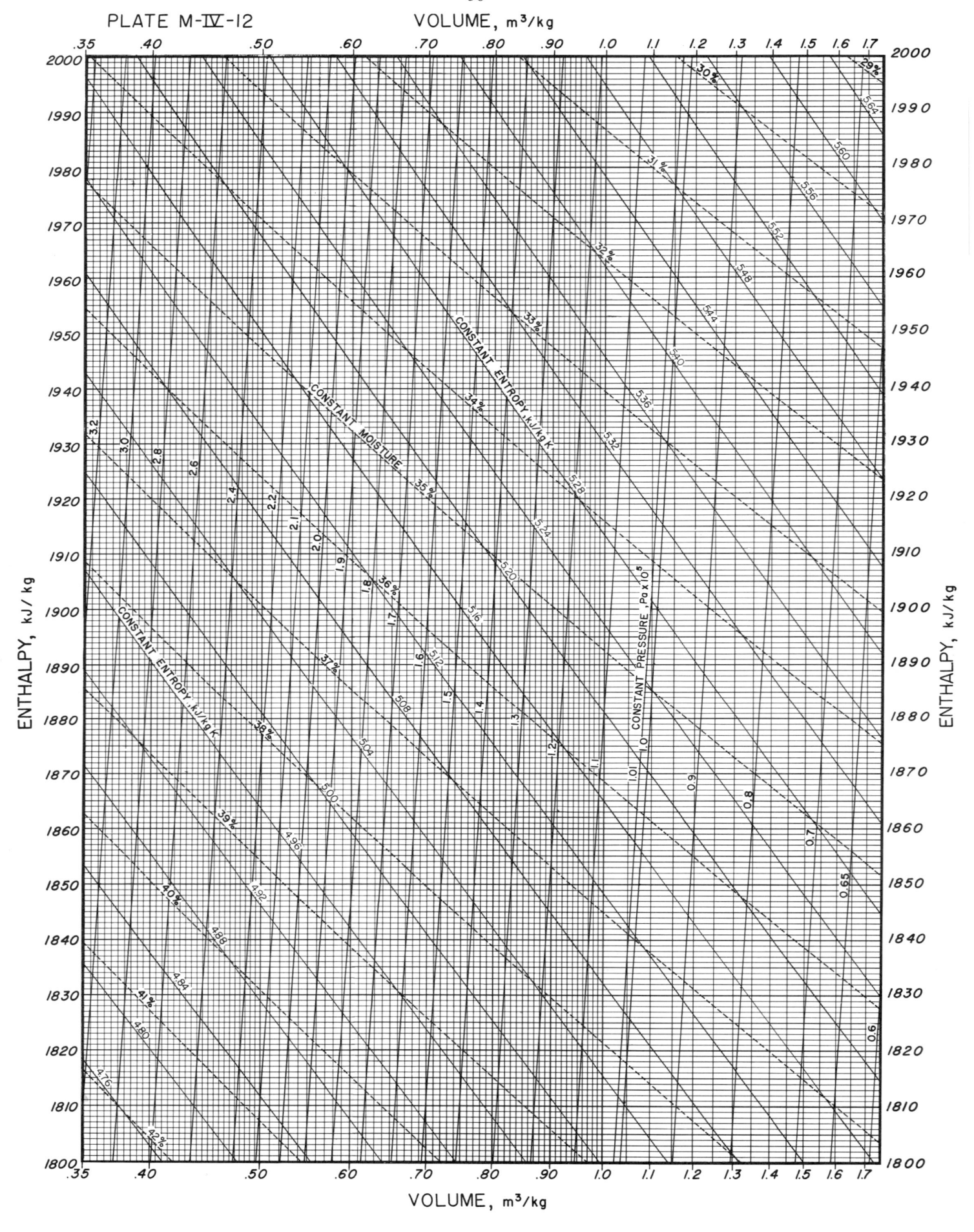

PLATE M-V-1

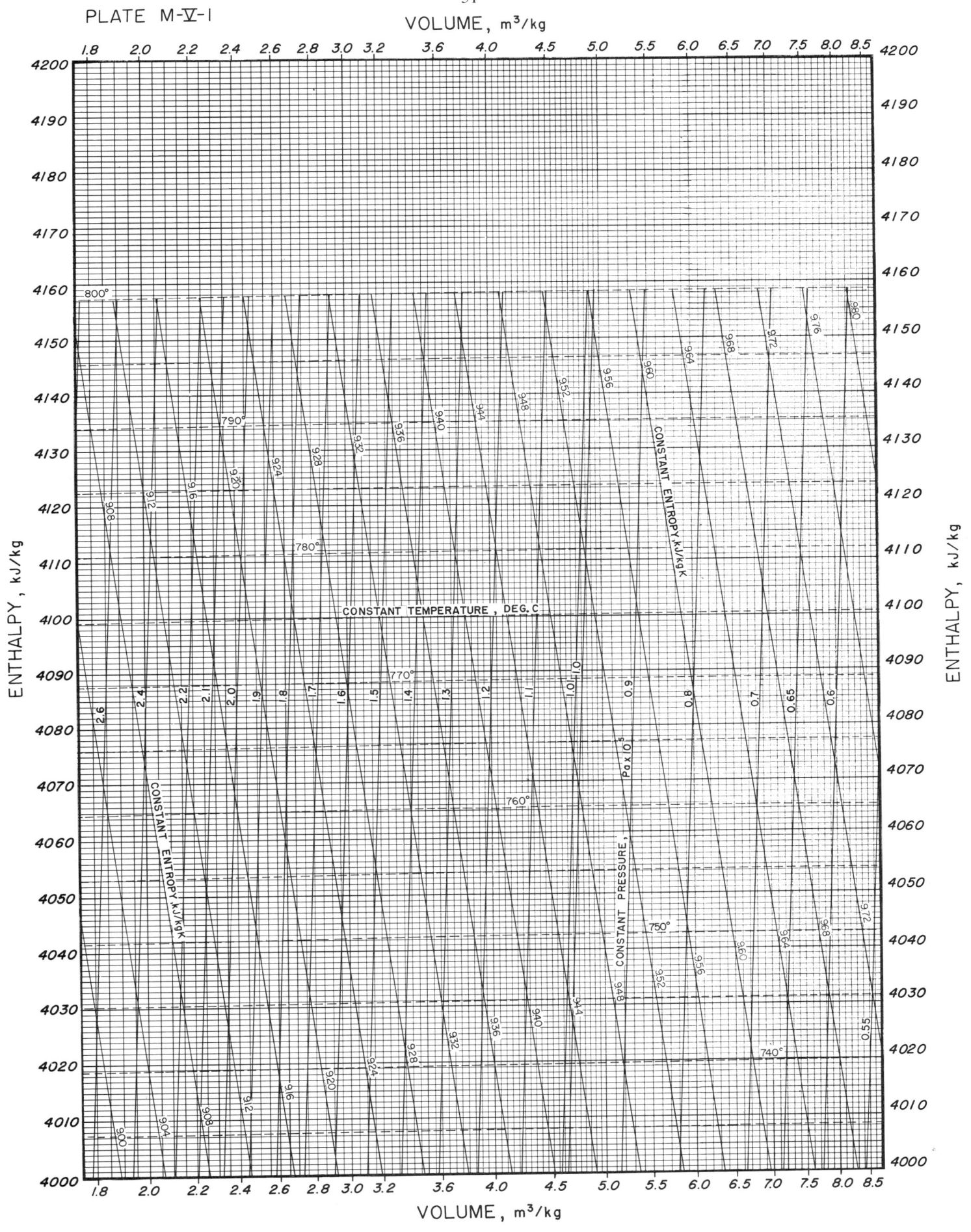

PLATE M-V-2

PLATE M-V-3

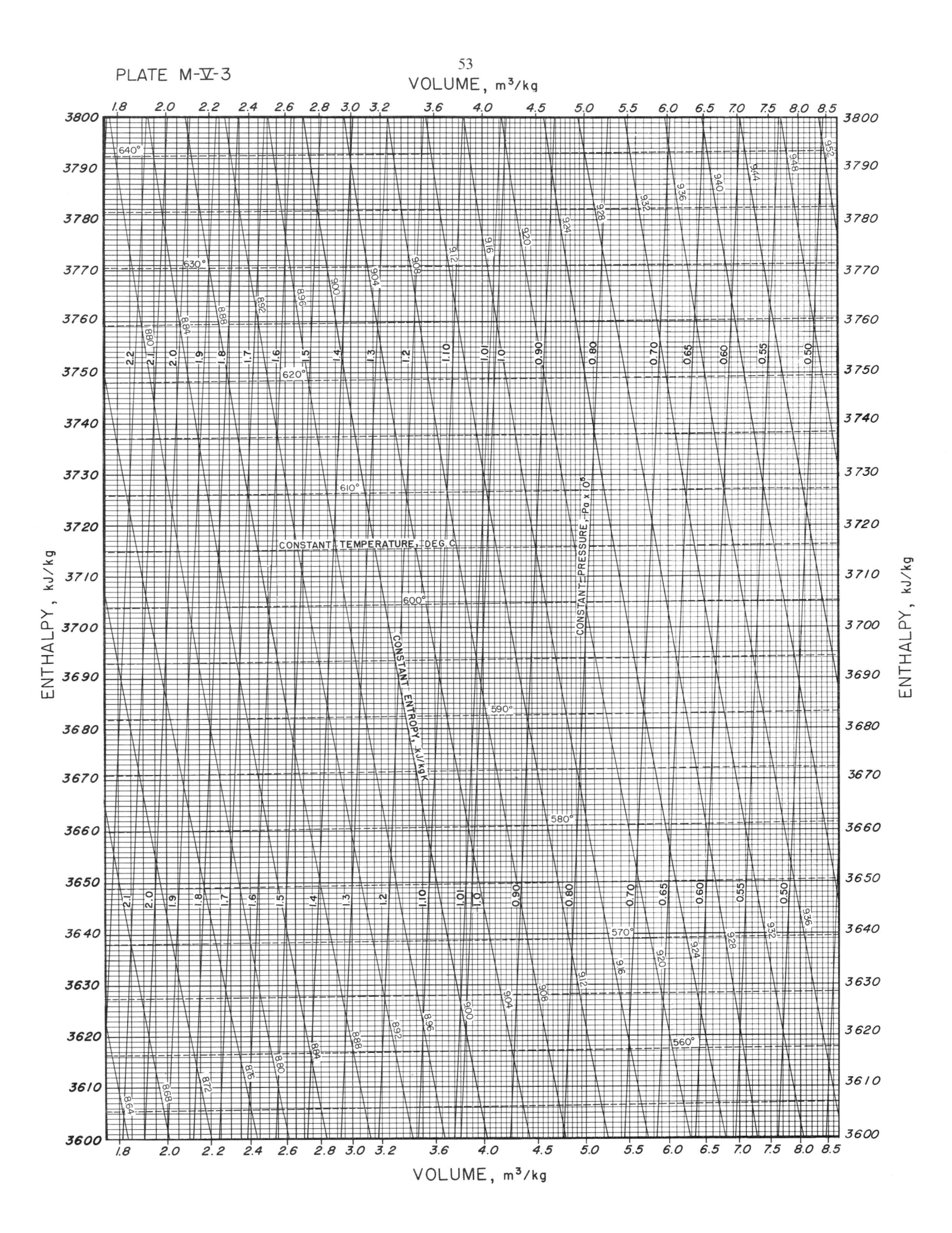

PLATE M-V-4

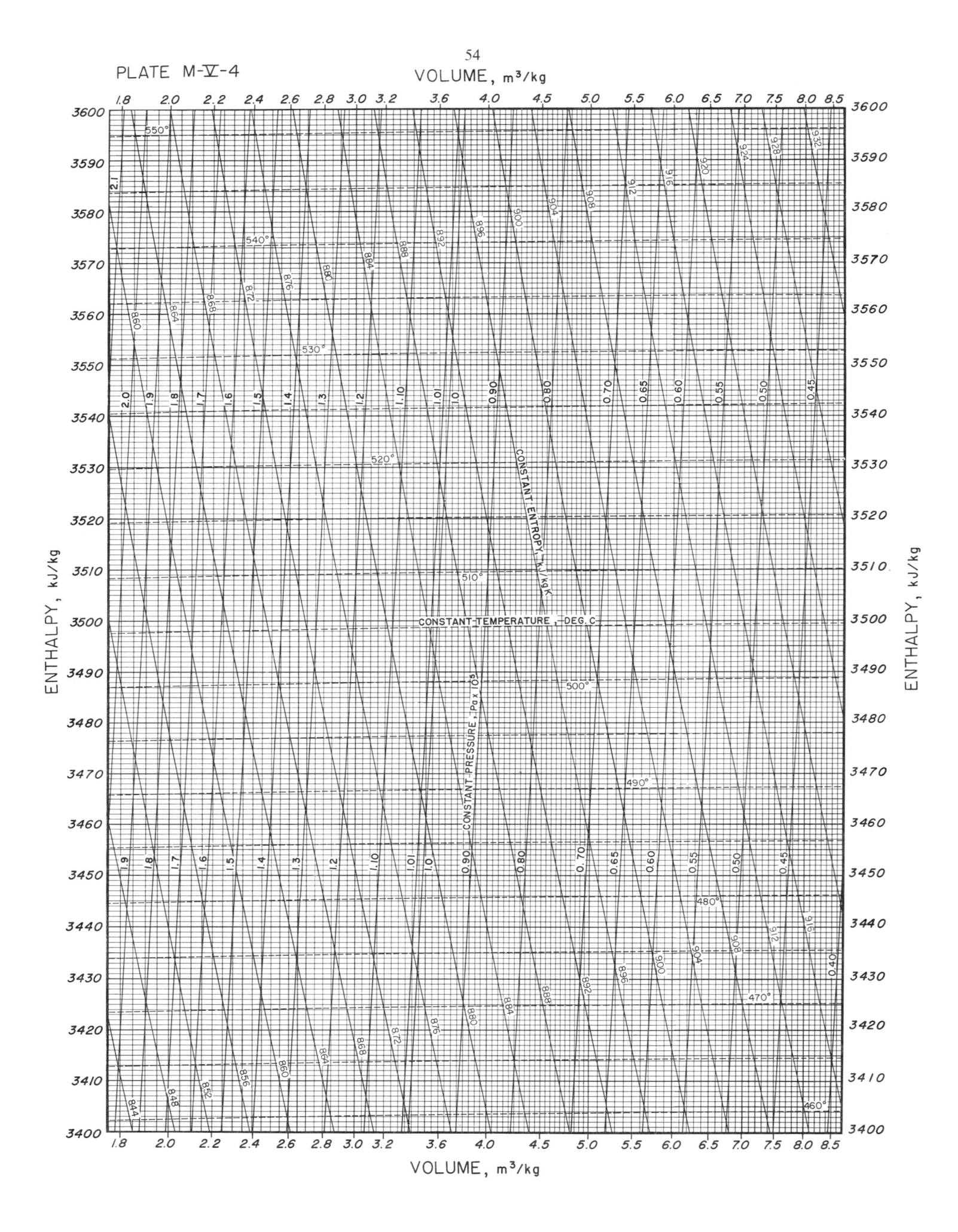

PLATE M-V-5

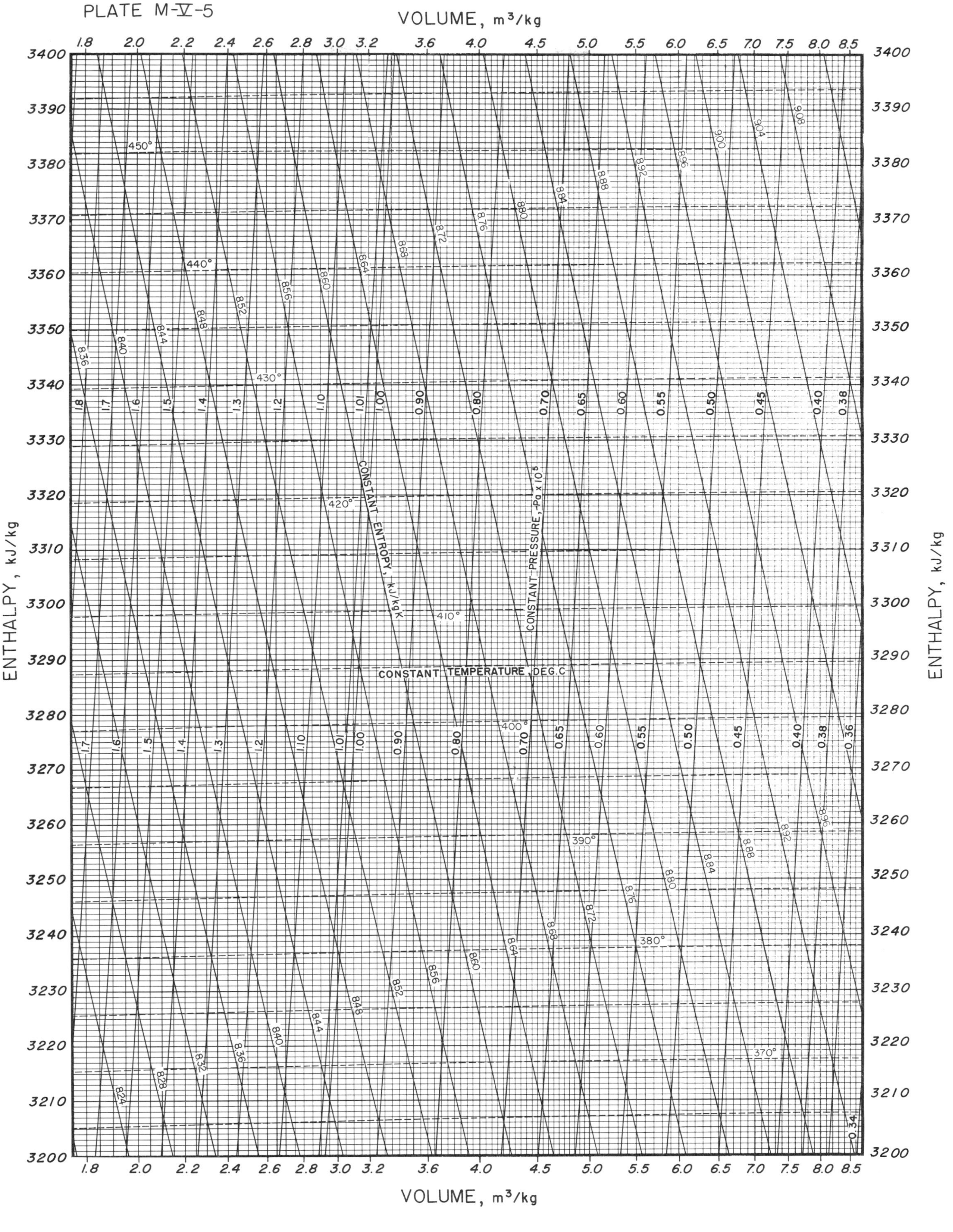
VOLUME, m³/kg
1.8 2.0 2.2 2.4 2.6 2.8 3.0 3.2 3.6 4.0 4.5 5.0 5.5 6.0 6.5 7.0 7.5 8.0 8.5
ENTHALPY, kJ/kg
3200 3210 3220 3230 3240 3250 3260 3270 3280 3290 3300 3310 3320 3330 3340 3350 3360 3370 3380 3390 3400
CONSTANT ENTROPY, kJ/kg K
CONSTANT PRESSURE, Pa x 10⁵
CONSTANT TEMPERATURE, DEG.C
370° 380° 390° 400° 410° 420° 430° 440° 450°
1.8 1.7 1.6 1.5 1.4 1.3 1.2 1.10 1.01 1.00 0.90 0.80 0.70 0.65 0.60 0.55 0.50 0.45 0.40 0.38 0.36 0.34
824 828 832 836 840 844 848 852 856 860 864 868 872 876 880 884 888 892 896 900 904 908

PLATE M-V-6

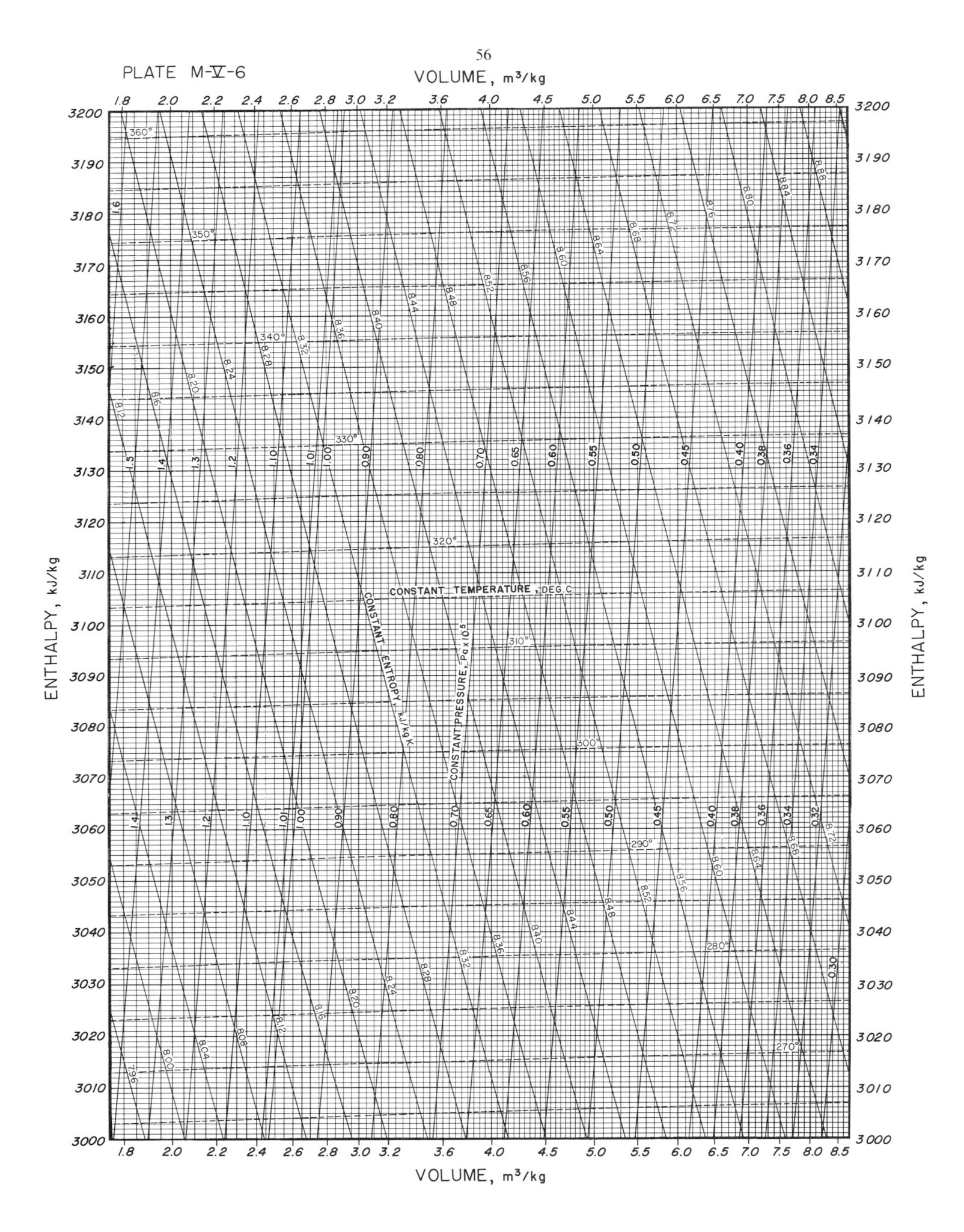

PLATE M-V-7

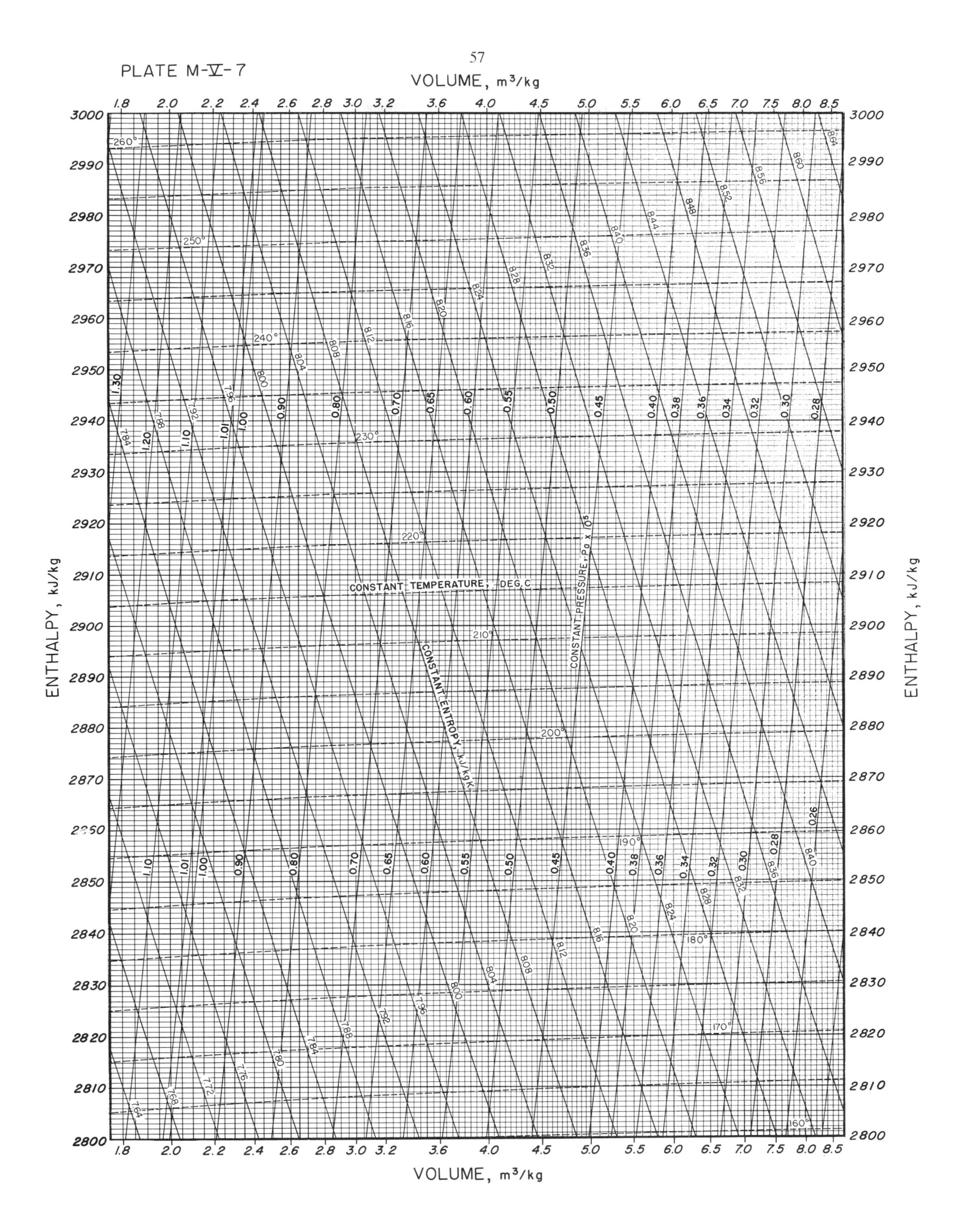

PLATE M-V-8

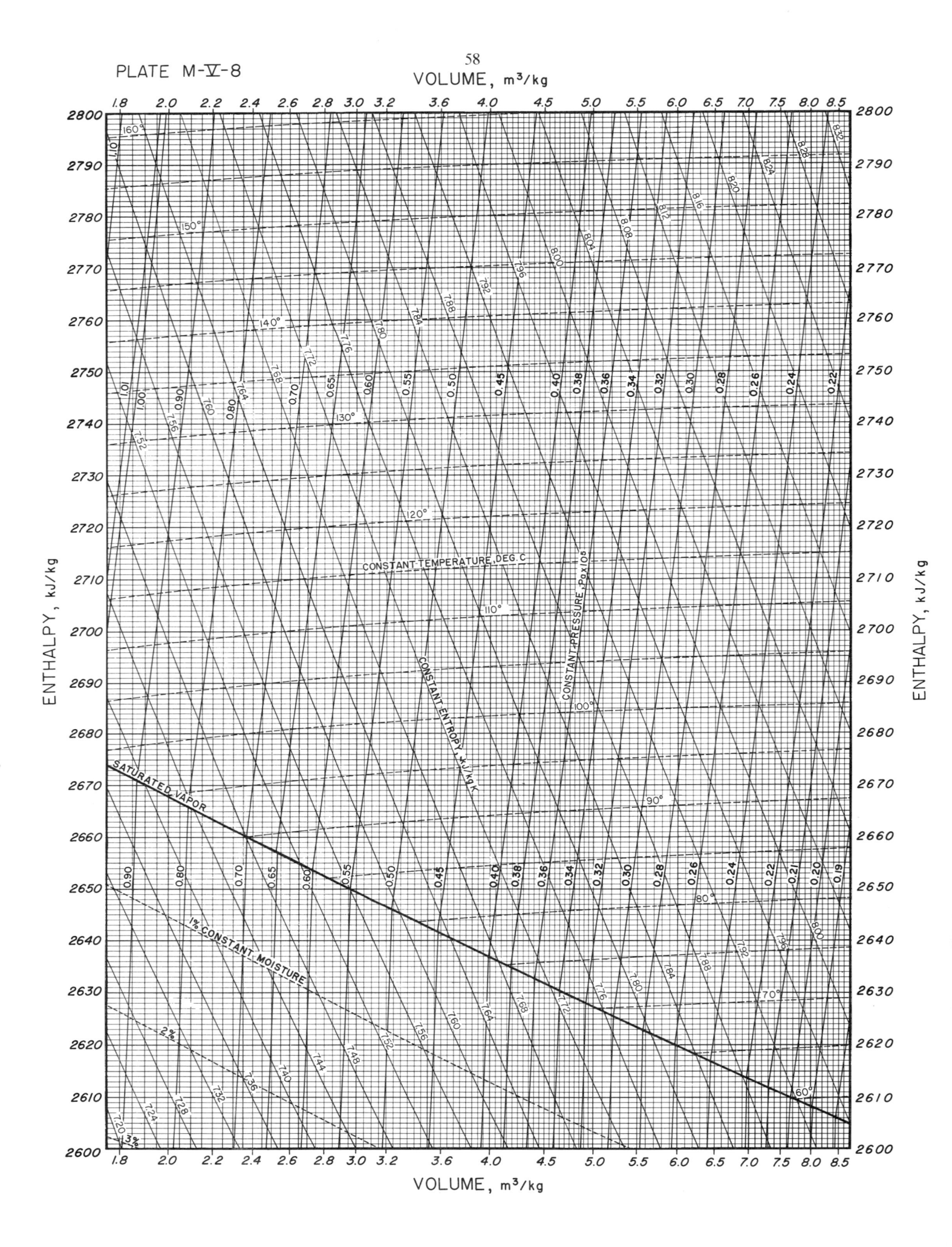

PLATE M-V-9

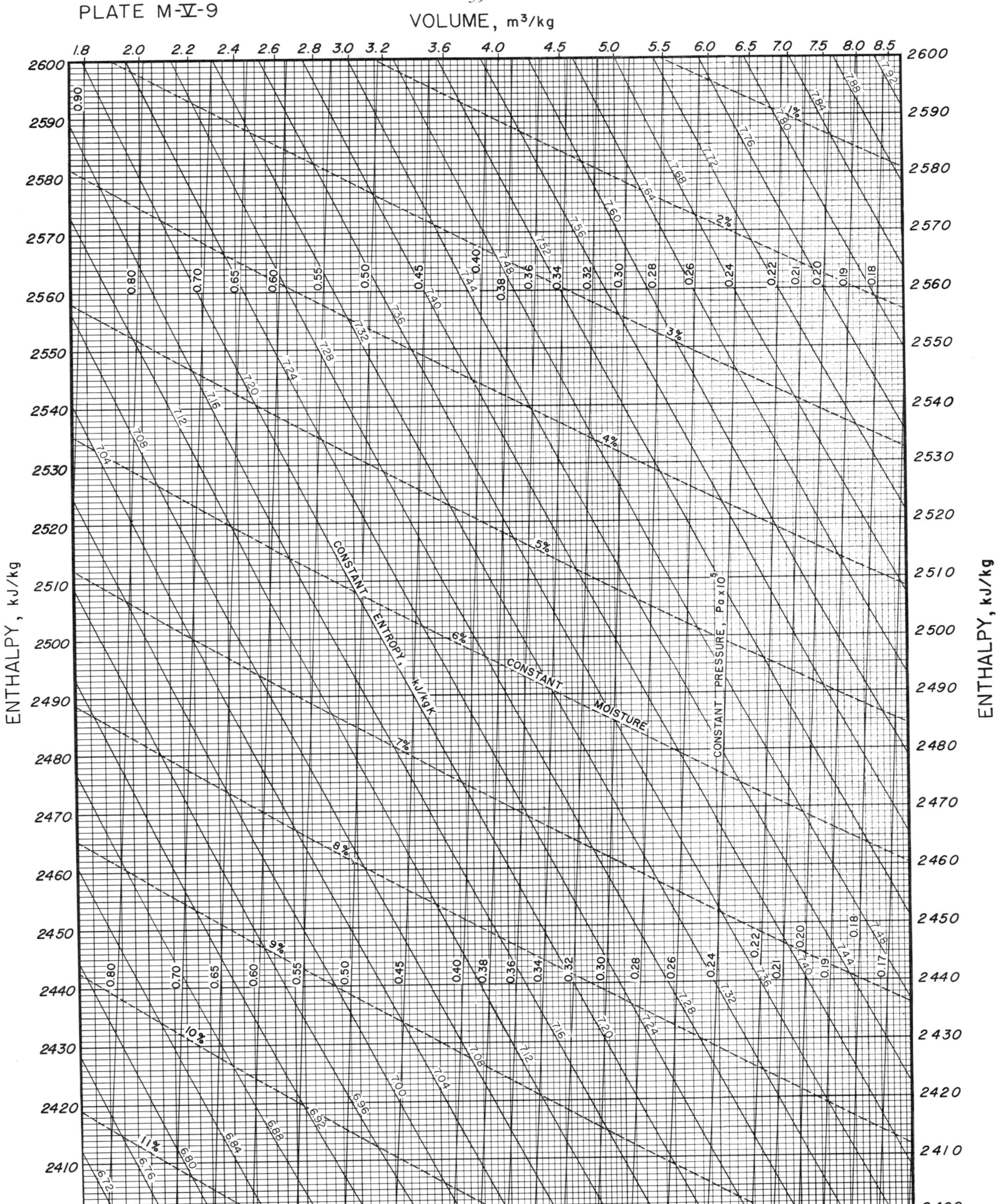
VOLUME, m³/kg
1.8 2.0 2.2 2.4 2.6 2.8 3.0 3.2 3.6 4.0 4.5 5.0 5.5 6.0 6.5 7.0 7.5 8.0 8.5
ENTHALPY, kJ/kg
2400 2410 2420 2430 2440 2450 2460 2470 2480 2490 2500 2510 2520 2530 2540 2550 2560 2570 2580 2590 2600
CONSTANT ENTROPY, kJ/kgK
CONSTANT MOISTURE
CONSTANT PRESSURE, Pa x 10⁵
1% 2% 3% 4% 5% 6% 7% 8% 9% 10% 11%
0.90 0.80 0.70 0.65 0.60 0.55 0.50 0.45 0.40 0.38 0.36 0.34 0.32 0.30 0.28 0.26 0.24 0.22 0.21 0.20 0.19 0.18 0.17
6.72 6.76 6.80 6.84 6.88 6.92 6.96 7.00 7.04 7.08 7.12 7.16 7.20 7.24 7.28 7.32 7.36 7.40 7.44 7.48 7.52 7.56 7.60 7.64 7.68 7.72 7.76 7.80 7.84 7.88 7.92

PLATE M-V-10

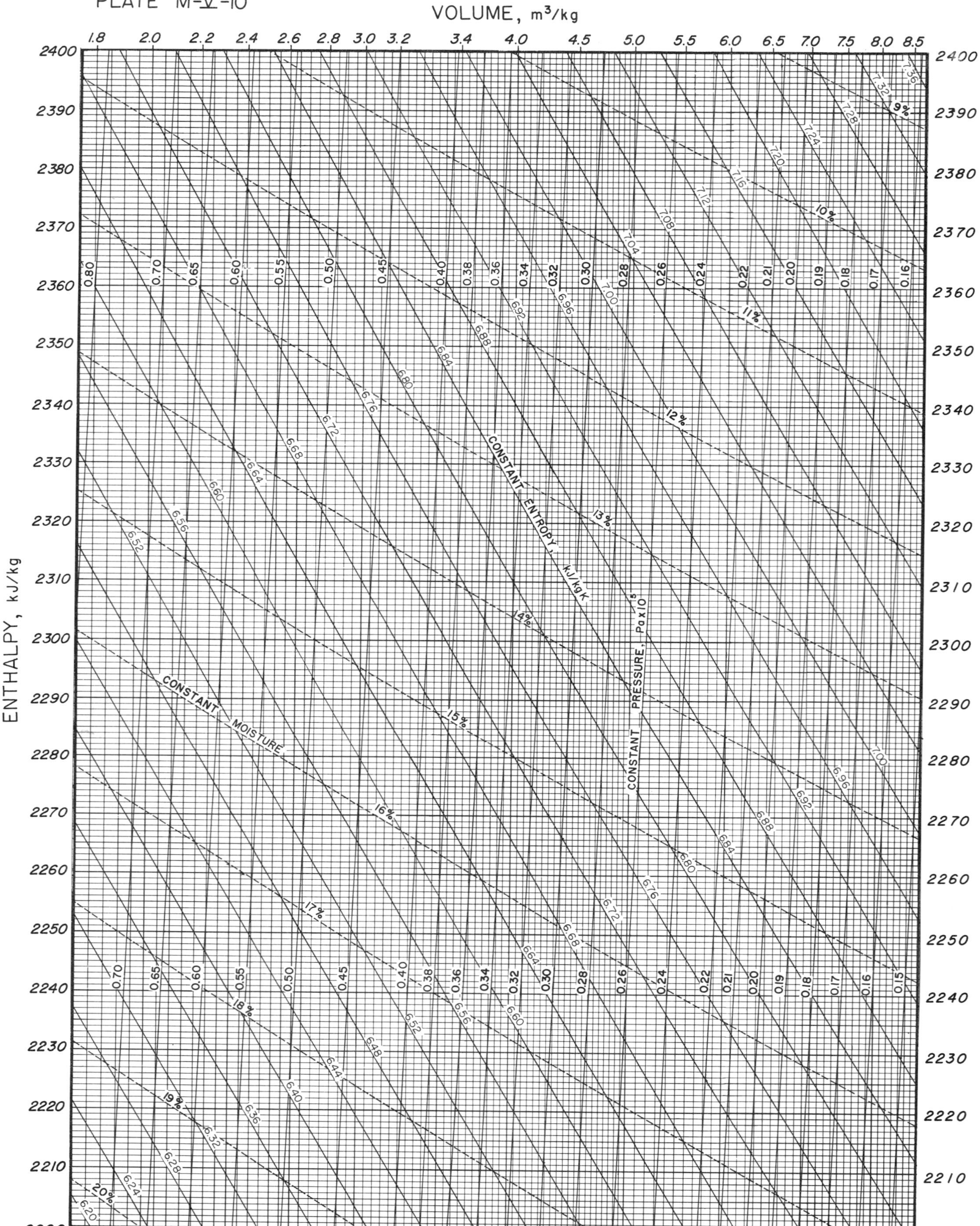
VOLUME, m³/kg
ENTHALPY, kJ/kg
CONSTANT ENTROPY, kJ/kg K
CONSTANT PRESSURE, Pa x 10⁵
CONSTANT MOISTURE
1.8
2.0
2.2
2.4
2.6
2.8
3.0
3.2
3.4
3.6
4.0
4.5
5.0
5.5
6.0
6.5
7.0
7.5
8.0
8.5
2200
2210
2220
2230
2240
2250
2260
2270
2280
2290
2300
2310
2320
2330
2340
2350
2360
2370
2380
2390
2400
9%
10%
11%
12%
13%
14%
15%
16%
17%
18%
19%
20%

PLATE M-V-11

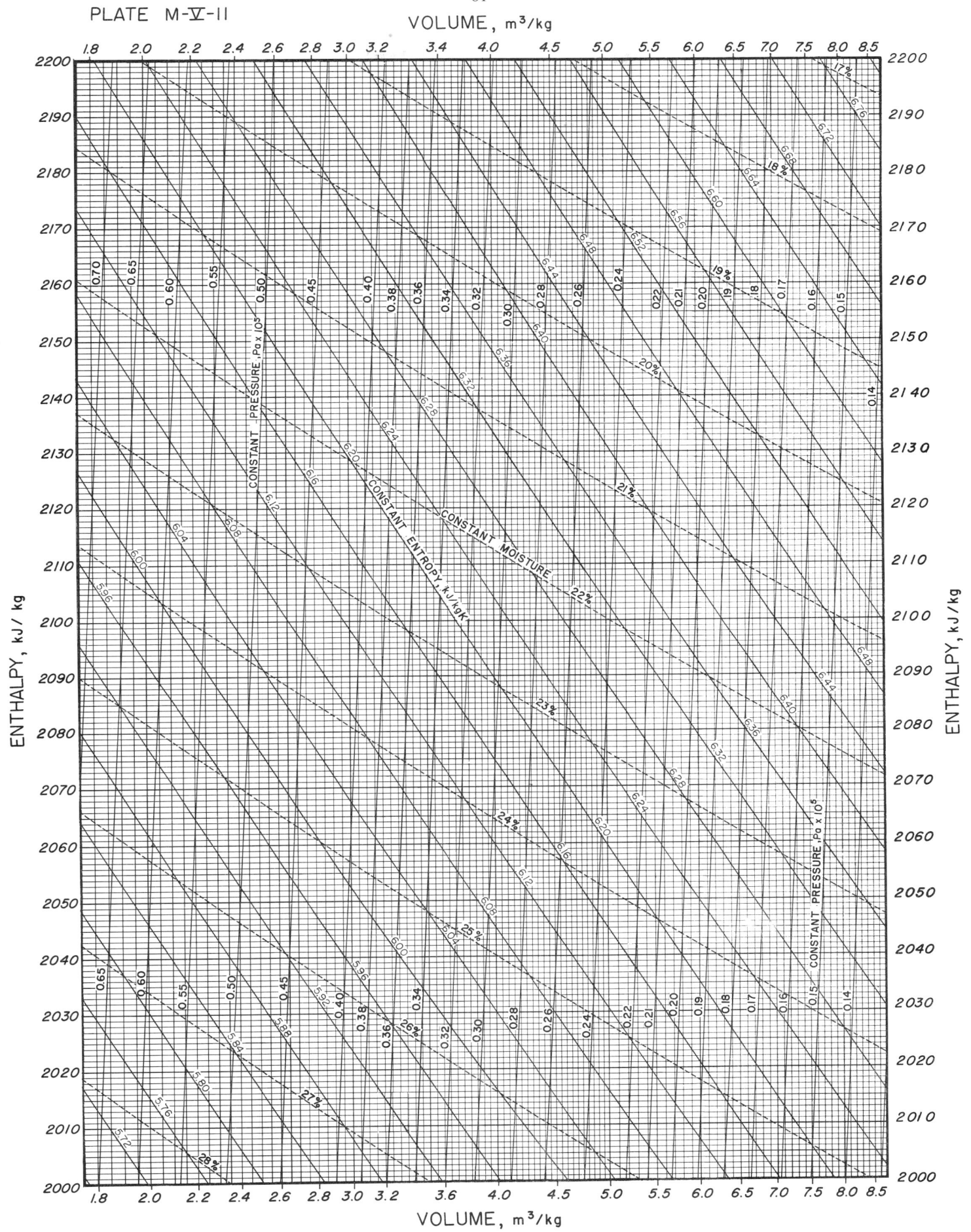

PLATE M-V-12

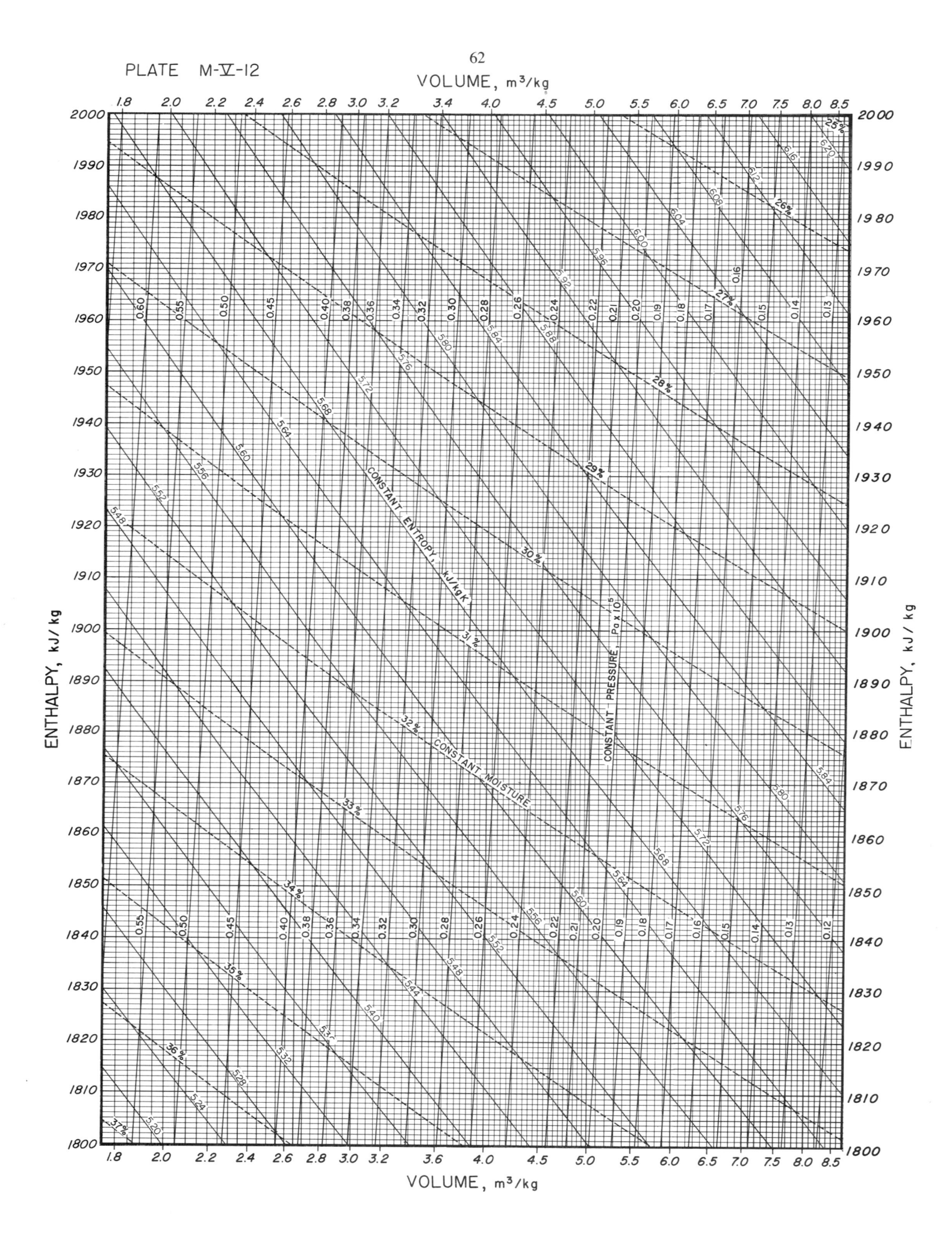

PLATE M-VI-I

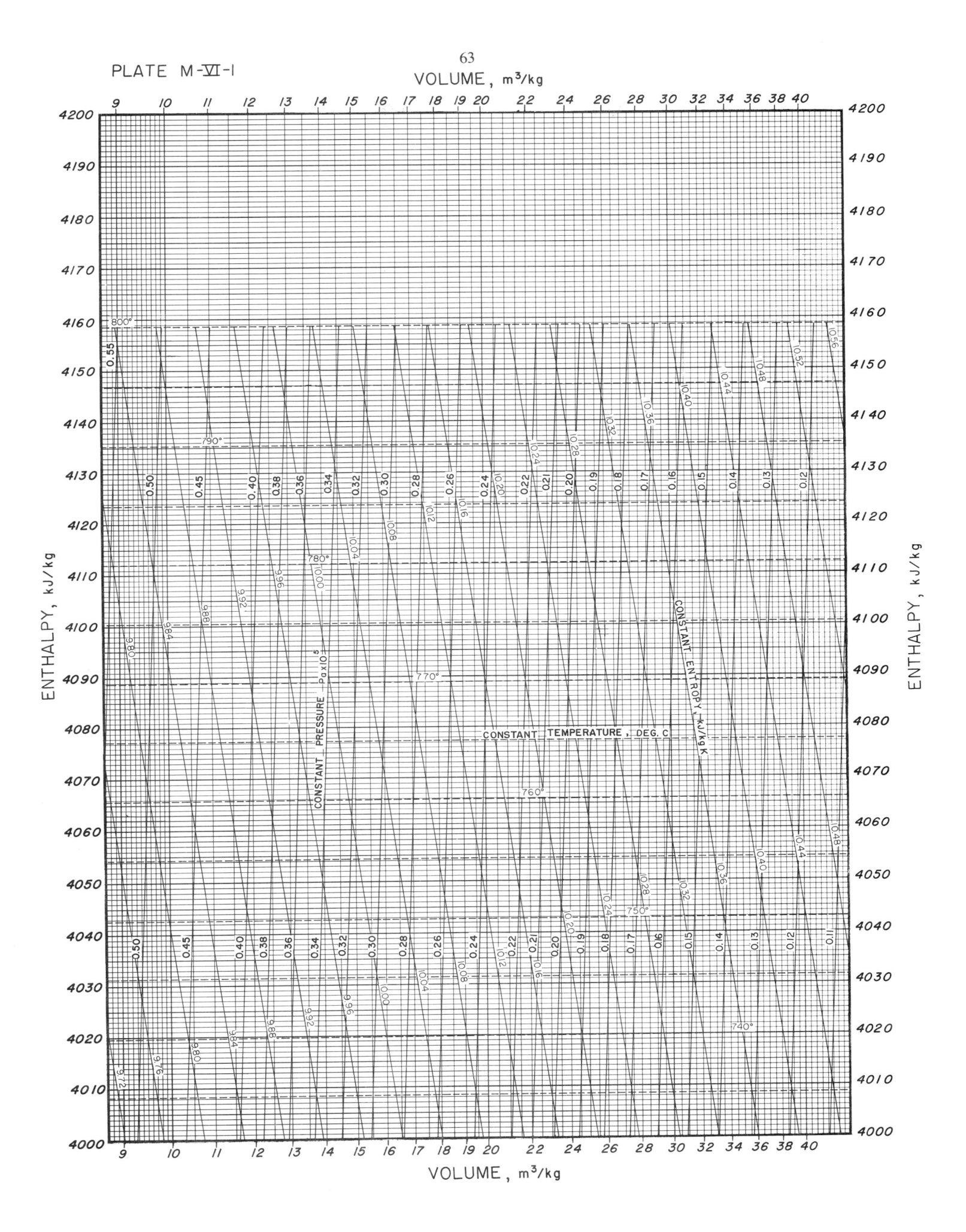

PLATE M-VI-2

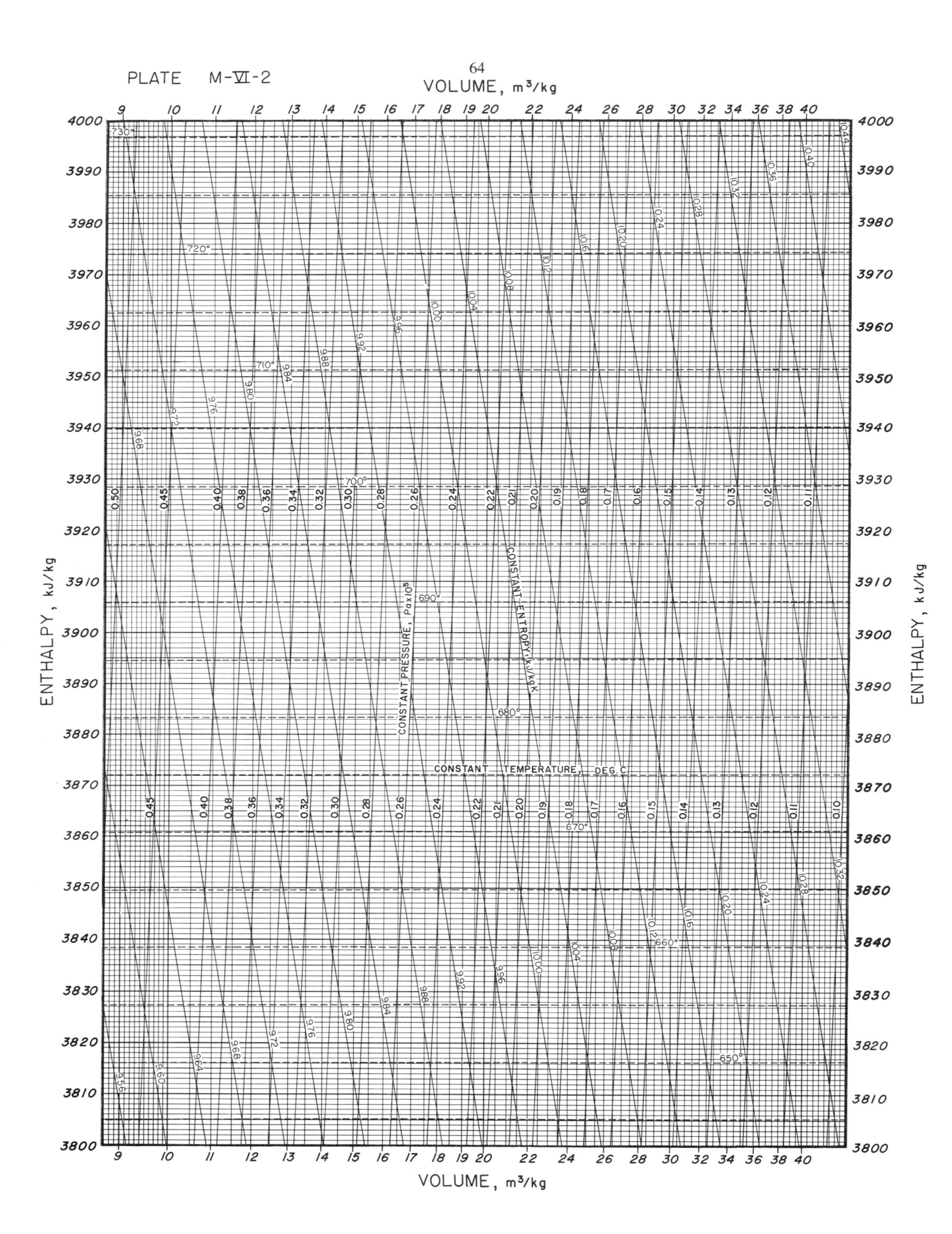

PLATE M-VI-3

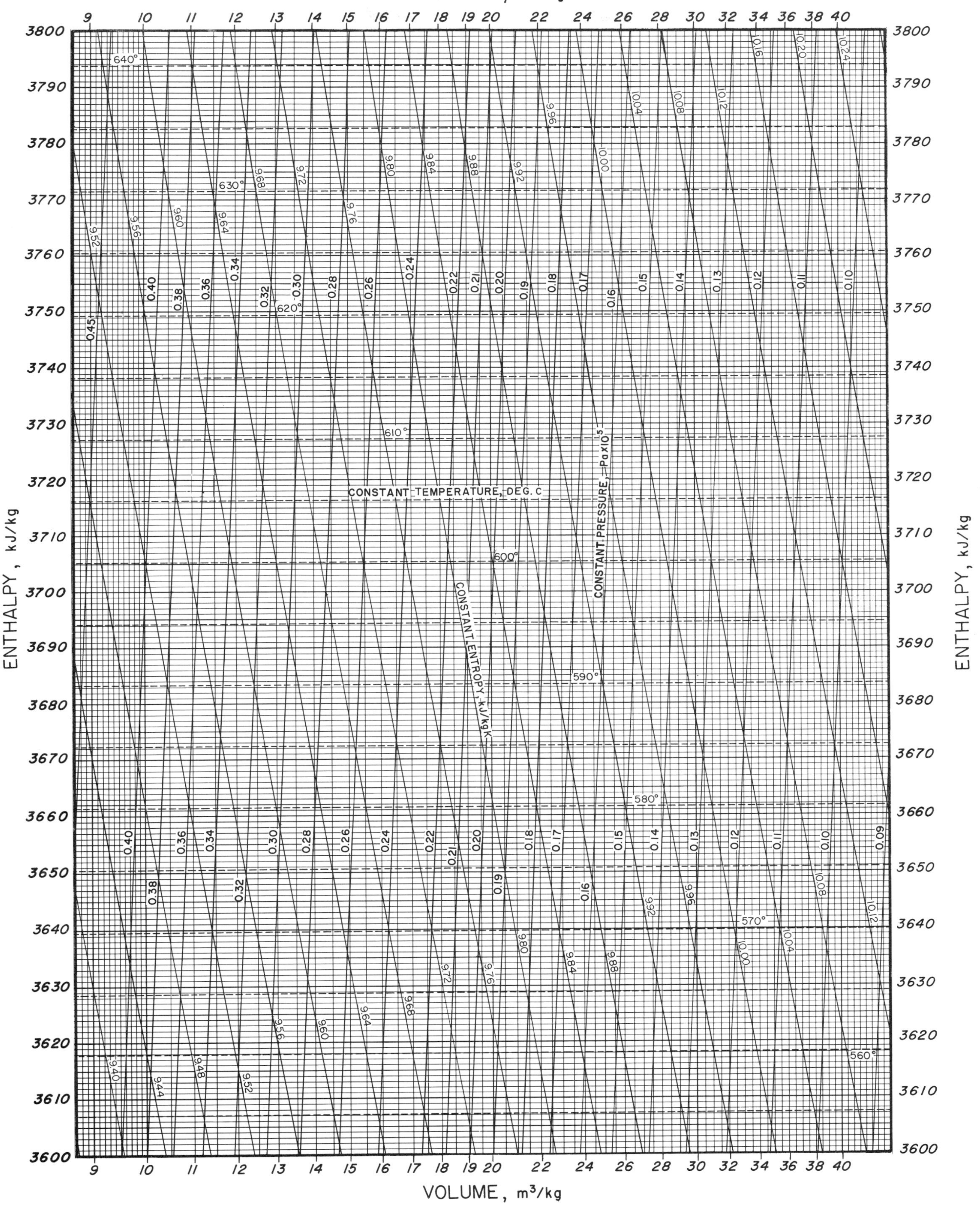
VOLUME, m³/kg
9 10 11 12 13 14 15 16 17 18 19 20 22 24 26 28 30 32 34 36 38 40
ENTHALPY, kJ/kg
3800 3790 3780 3770 3760 3750 3740 3730 3720 3710 3700 3690 3680 3670 3660 3650 3640 3630 3620 3610 3600
CONSTANT TEMPERATURE, DEG. C
CONSTANT ENTROPY, kJ/kg K
CONSTANT PRESSURE, Pa x 10⁻⁵
640° 630° 620° 610° 600° 590° 580° 570° 560°
0.45 0.40 0.38 0.36 0.34 0.32 0.30 0.28 0.26 0.24 0.22 0.21 0.20 0.19 0.18 0.17 0.16 0.15 0.14 0.13 0.12 0.11 0.10 0.09
9.40 9.44 9.48 9.52 9.56 9.60 9.64 9.68 9.72 9.76 9.80 9.84 9.88 9.92 9.96 10.00 10.04 10.08 10.12 10.16 10.20 10.24

PLATE M-VI-4

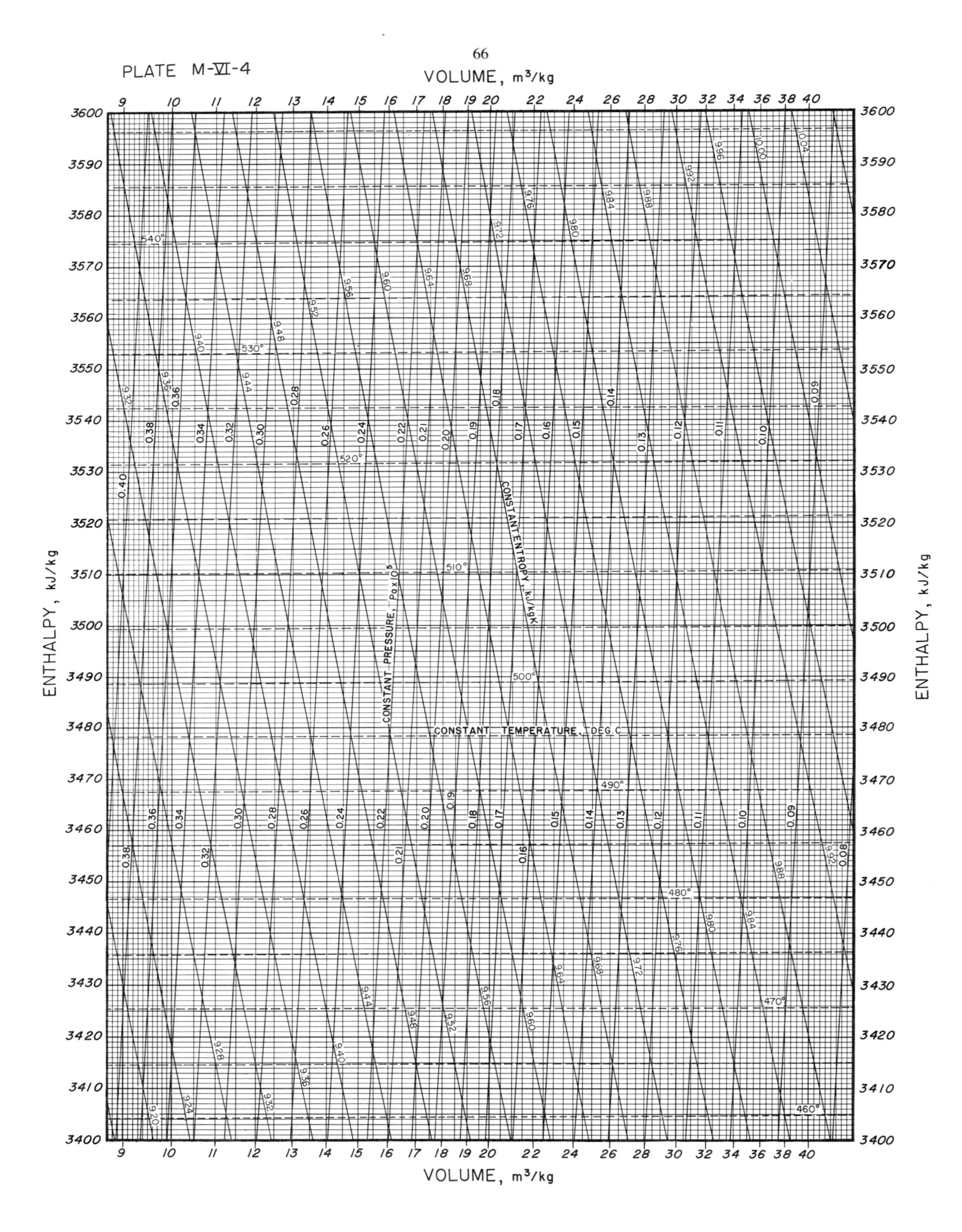

PLATE M-VI-5

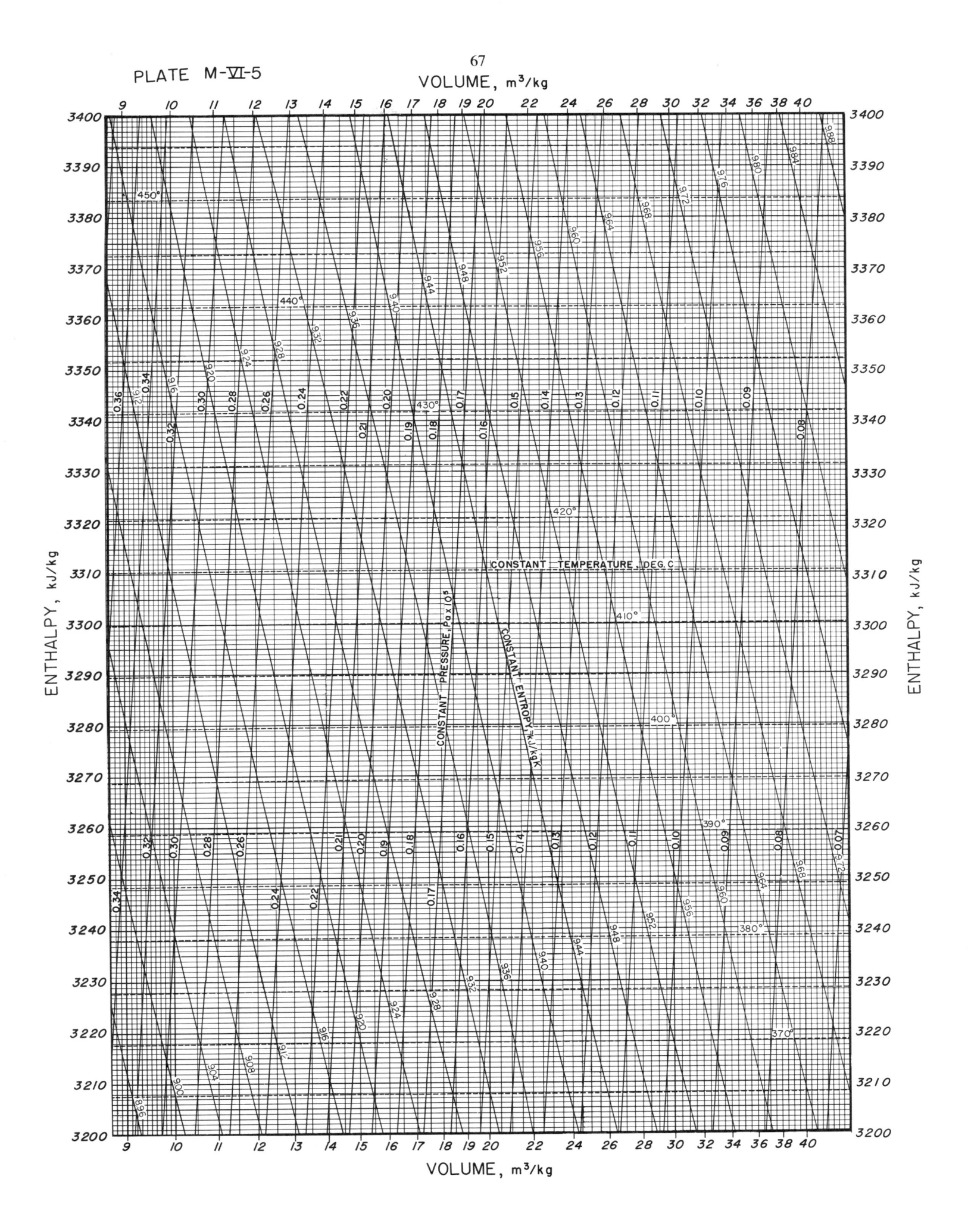

PLATE M-VI-6

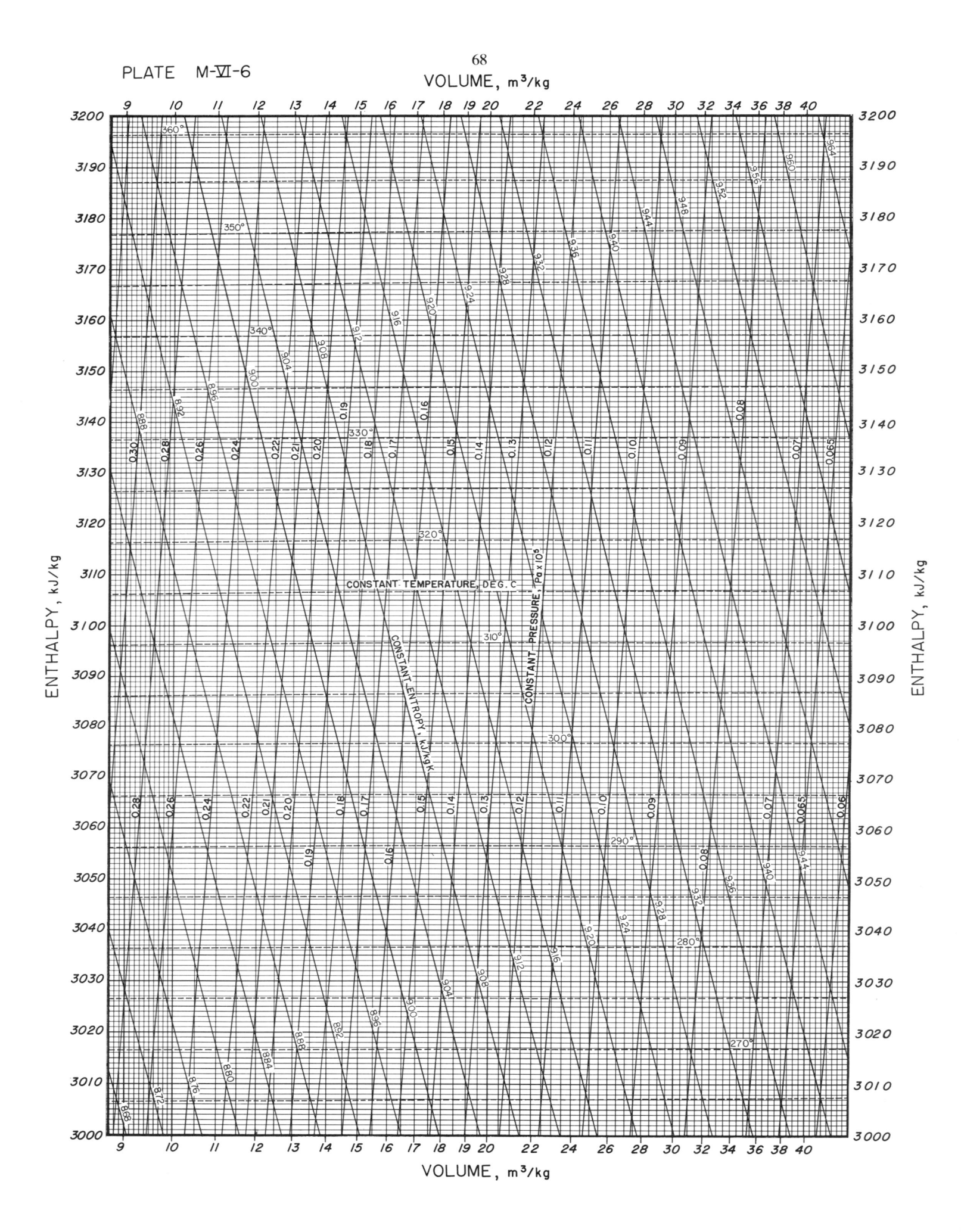

PLATE M-VI-7

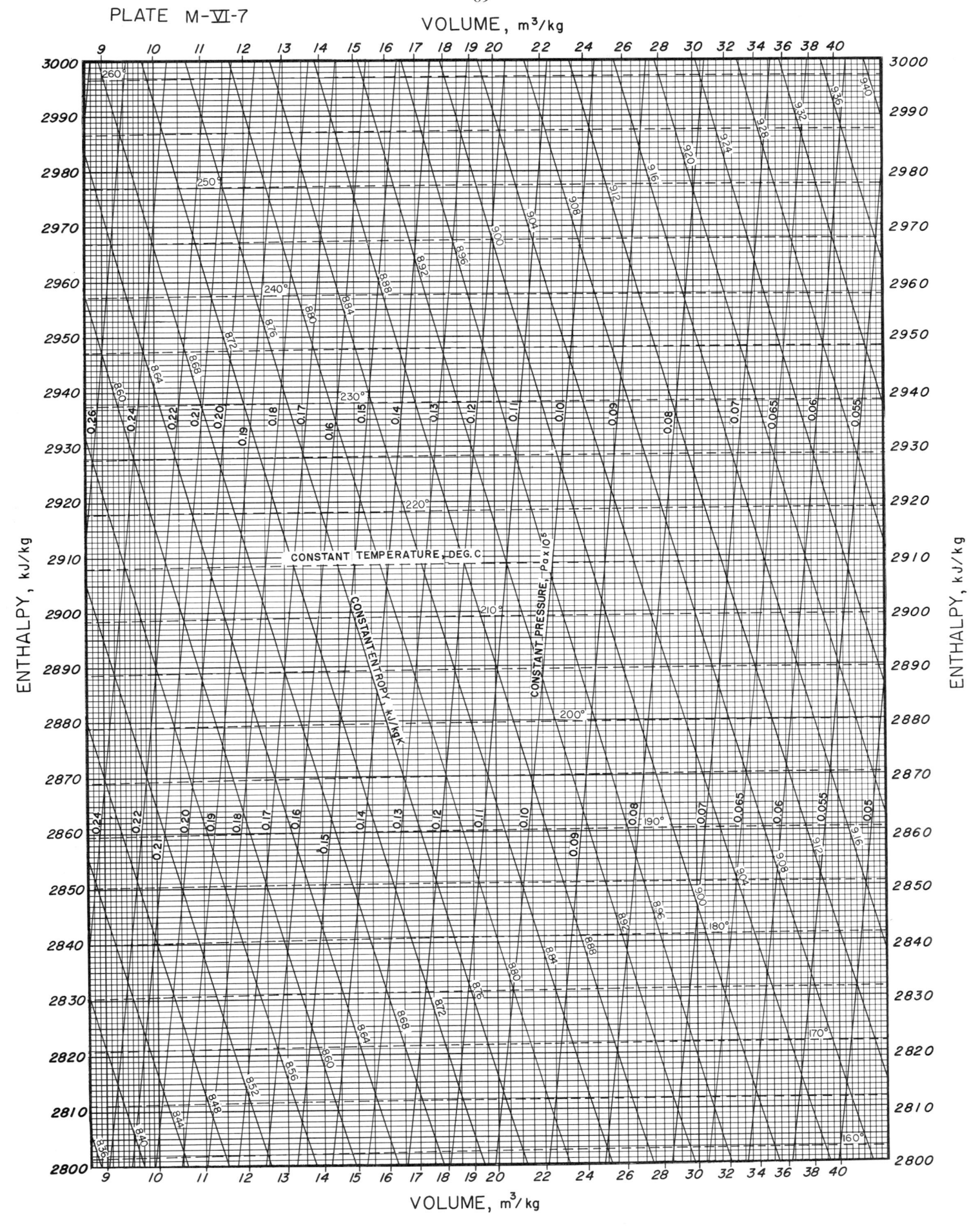

PLATE M-VI-8

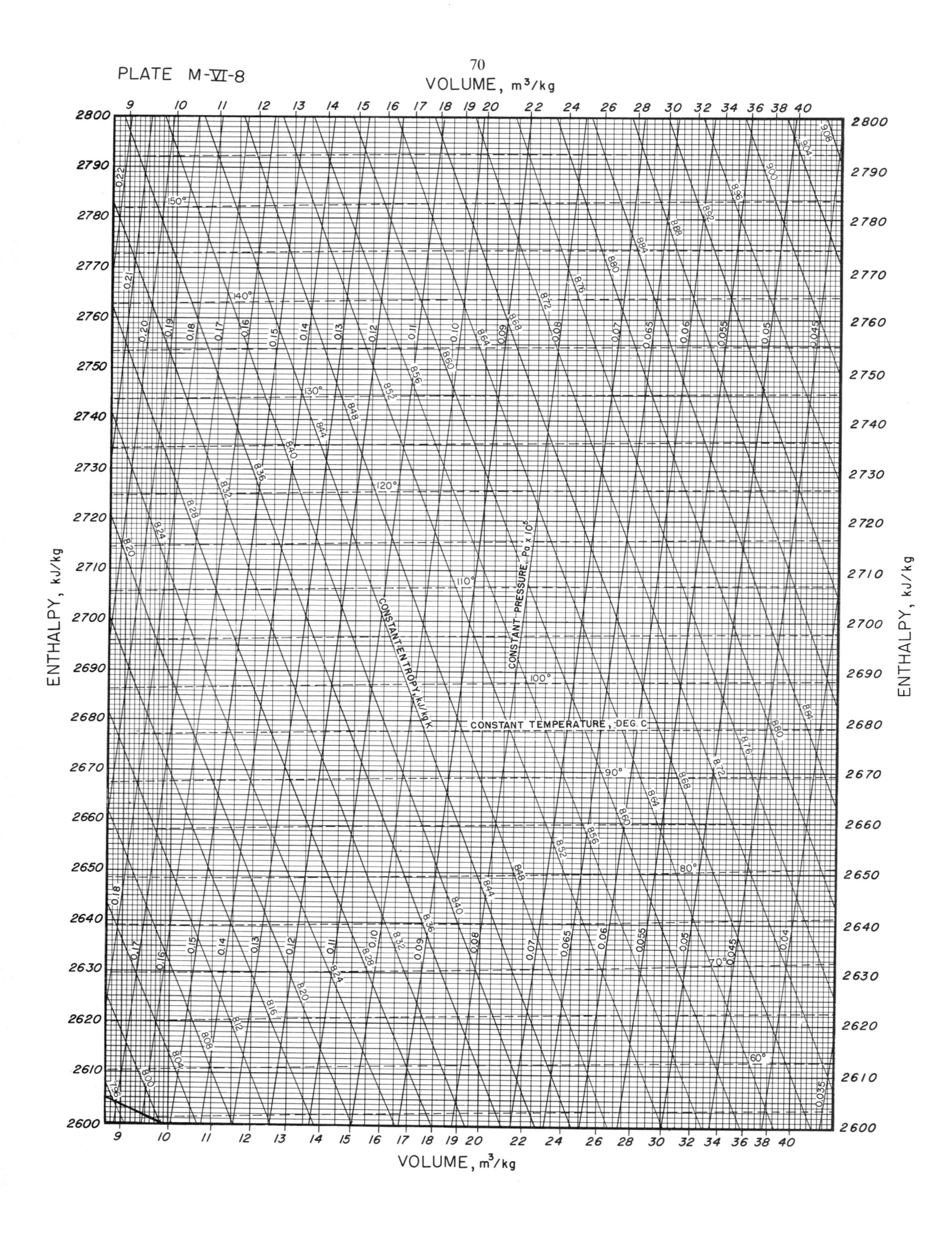

PLATE M-VI-9

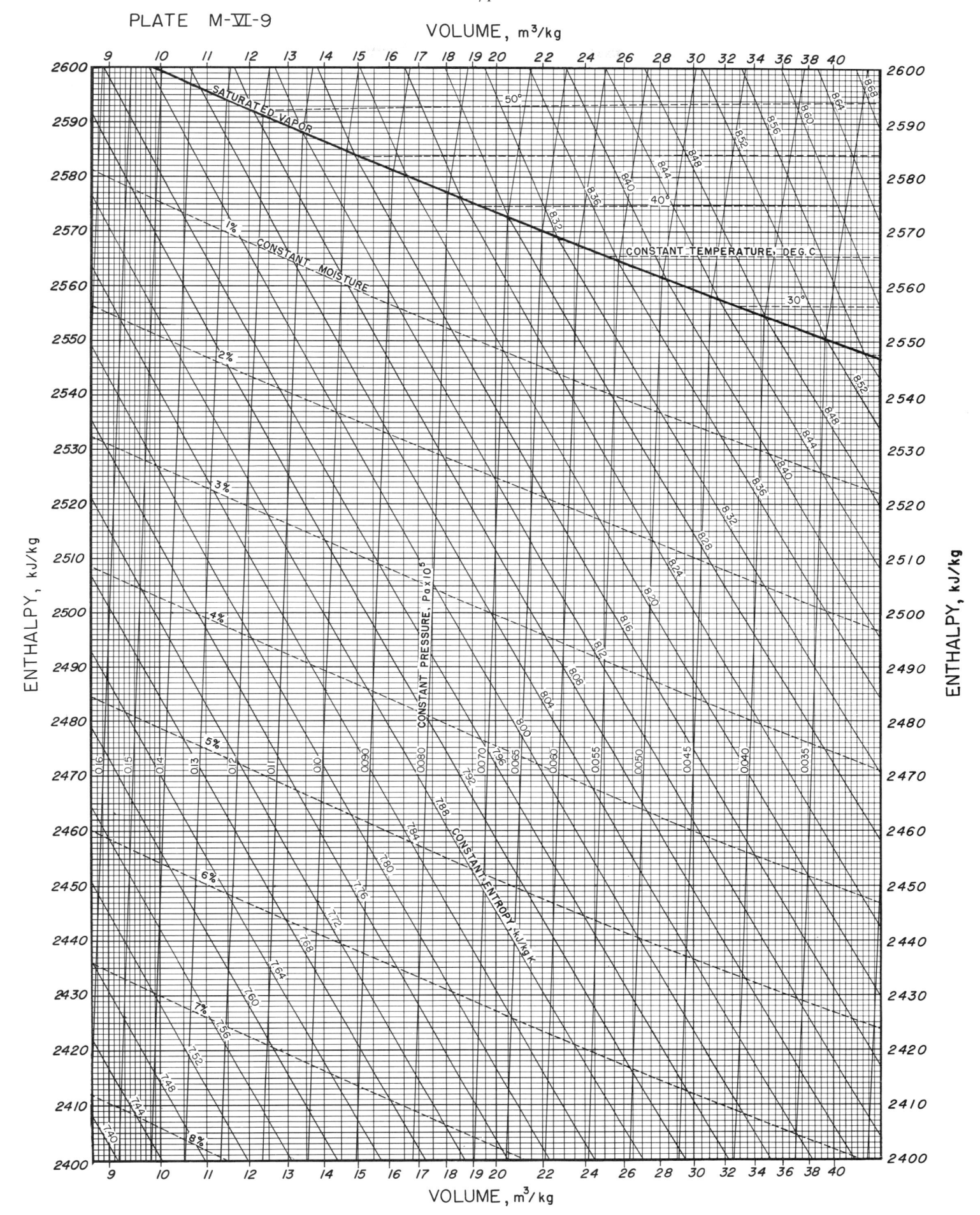

PLATE M-VI-10

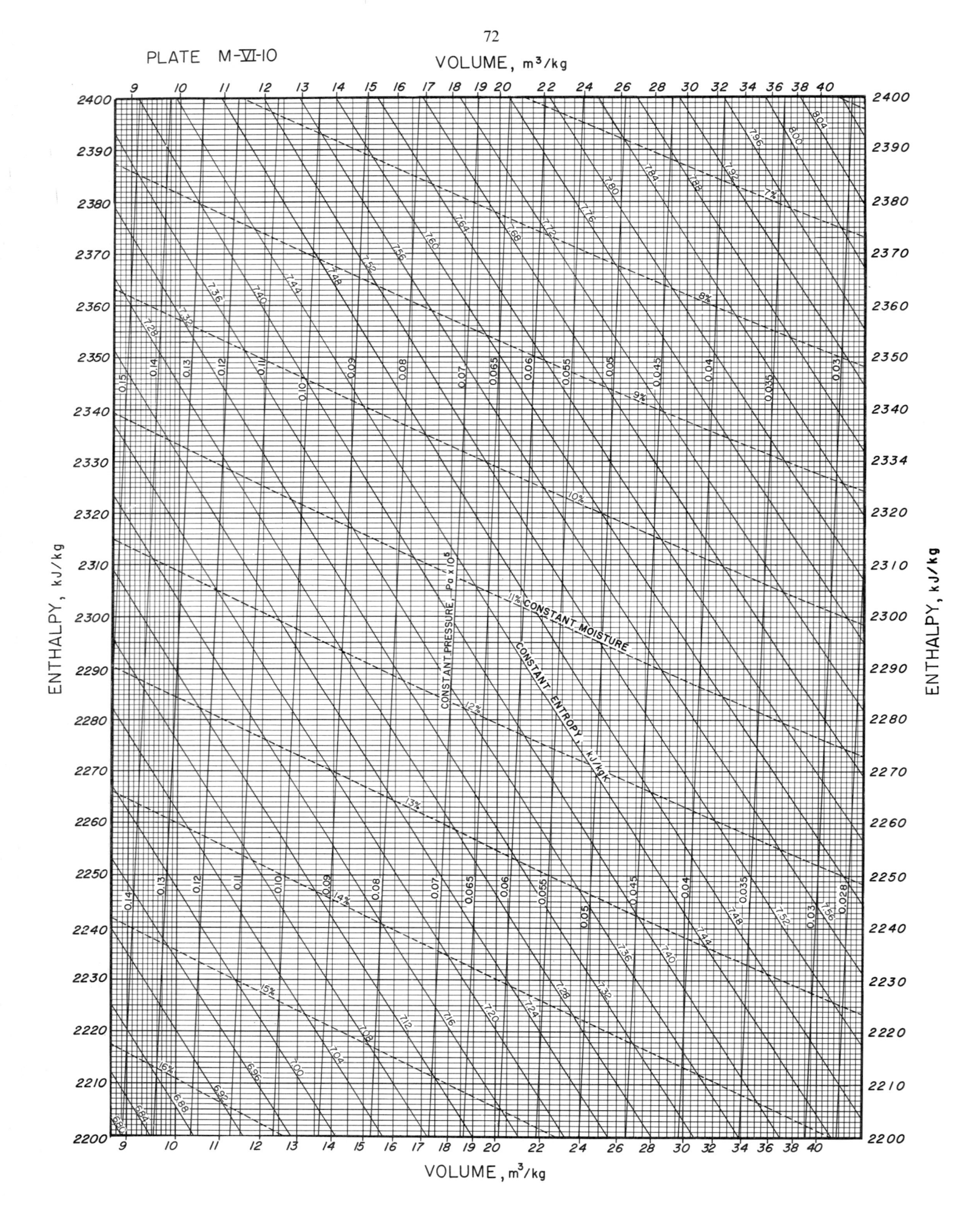

PLATE M-VI-II

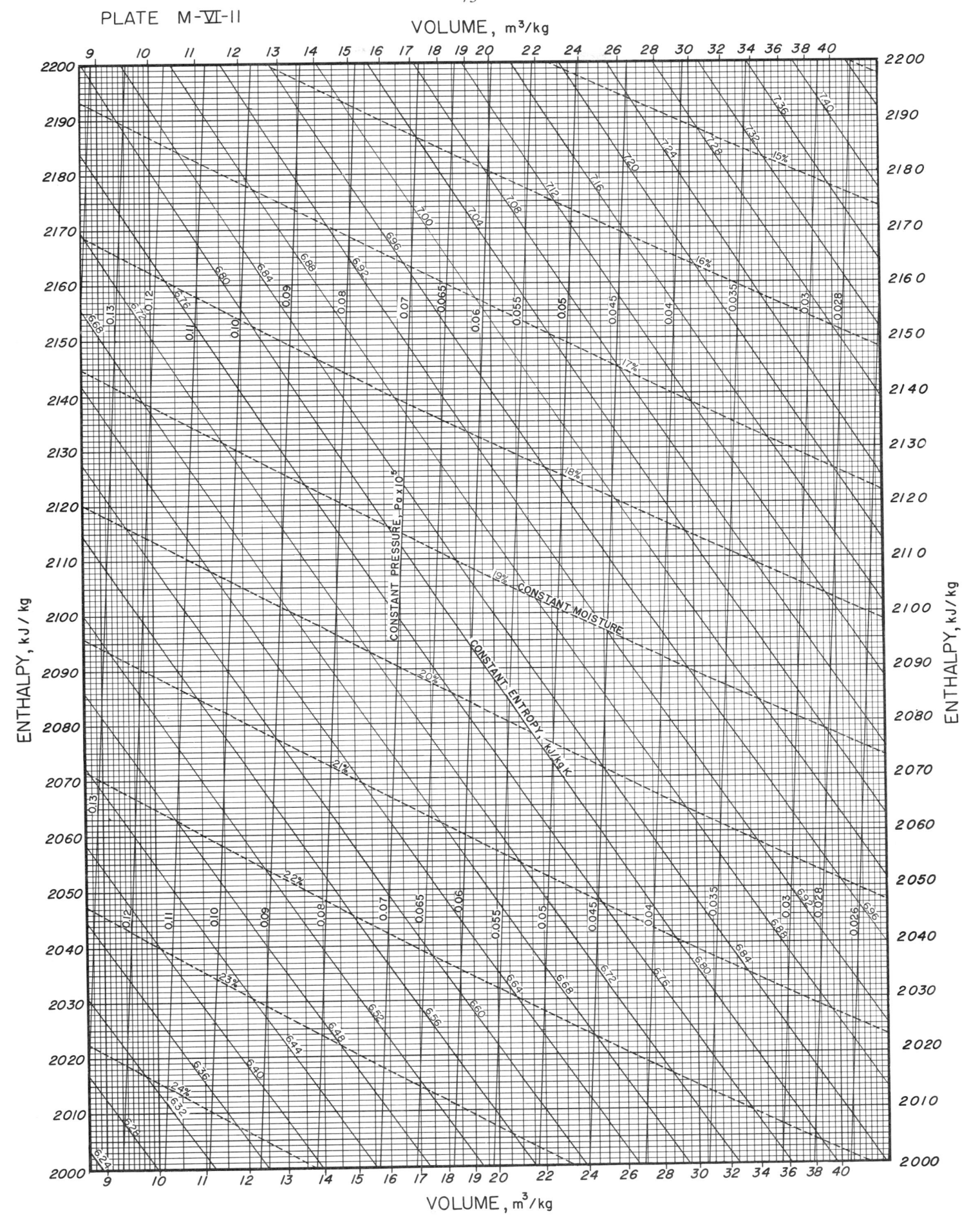

PLATE M-VI-12

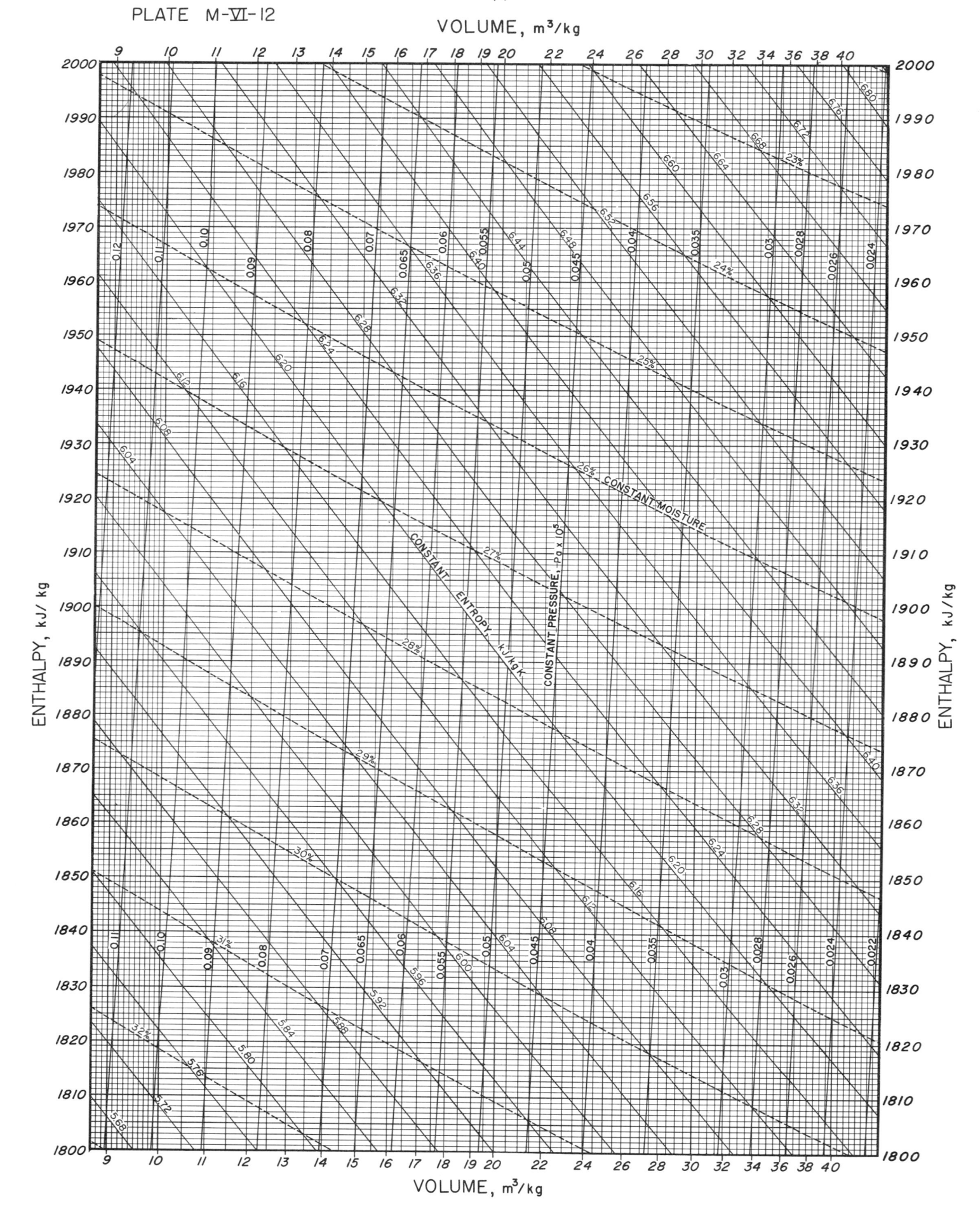

PLATE M-VII-I

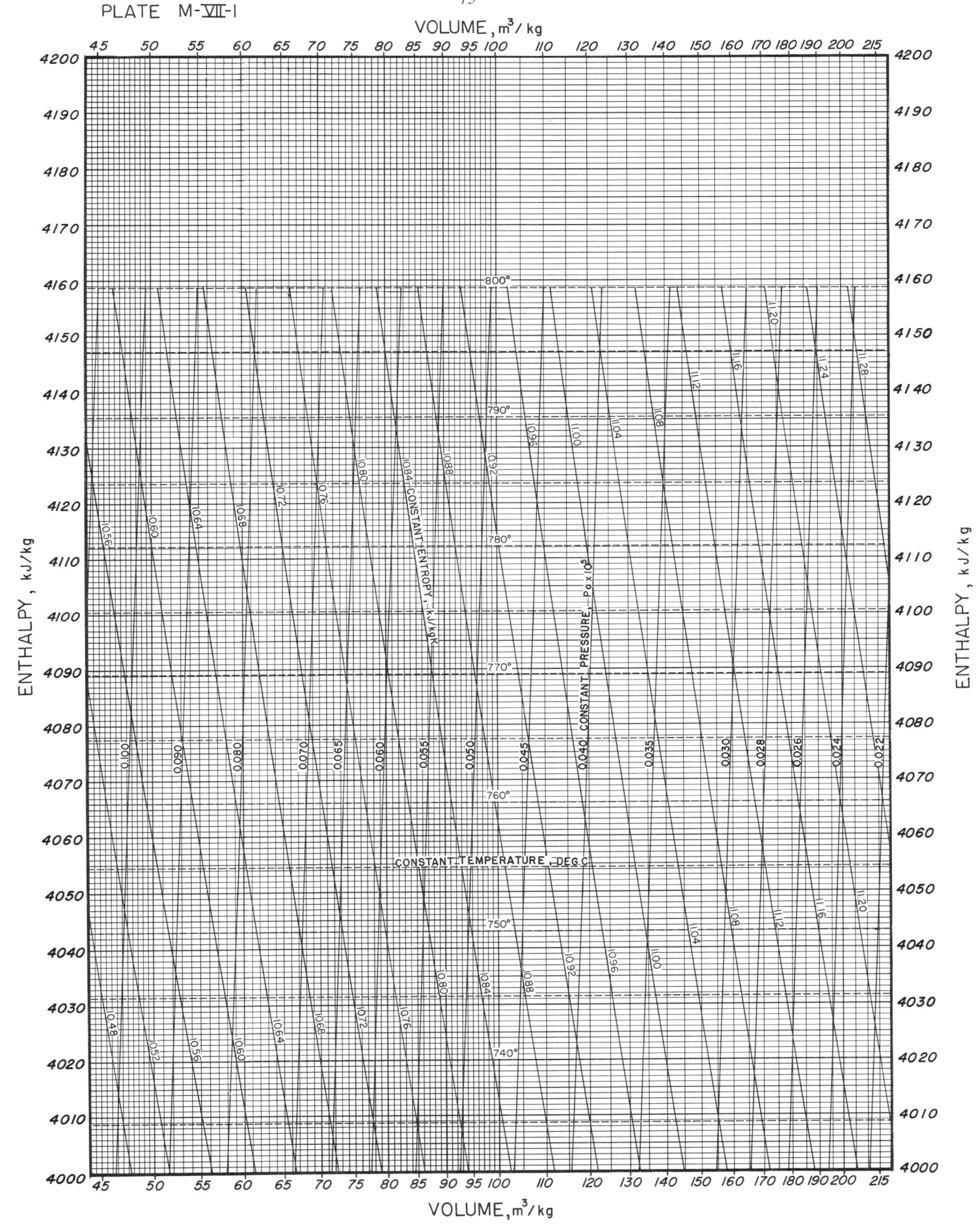

PLATE M-VII-2

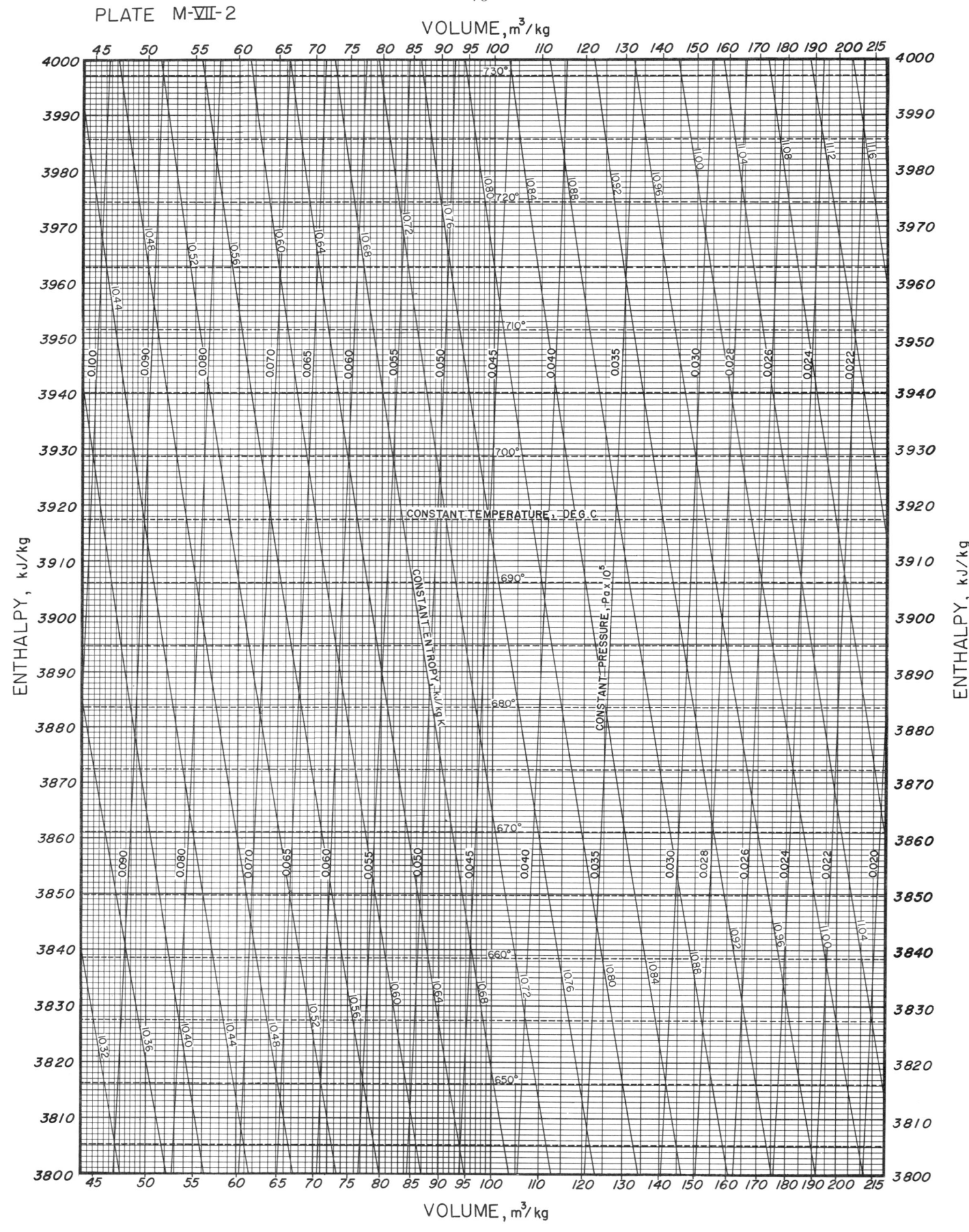

PLATE M-VII-3

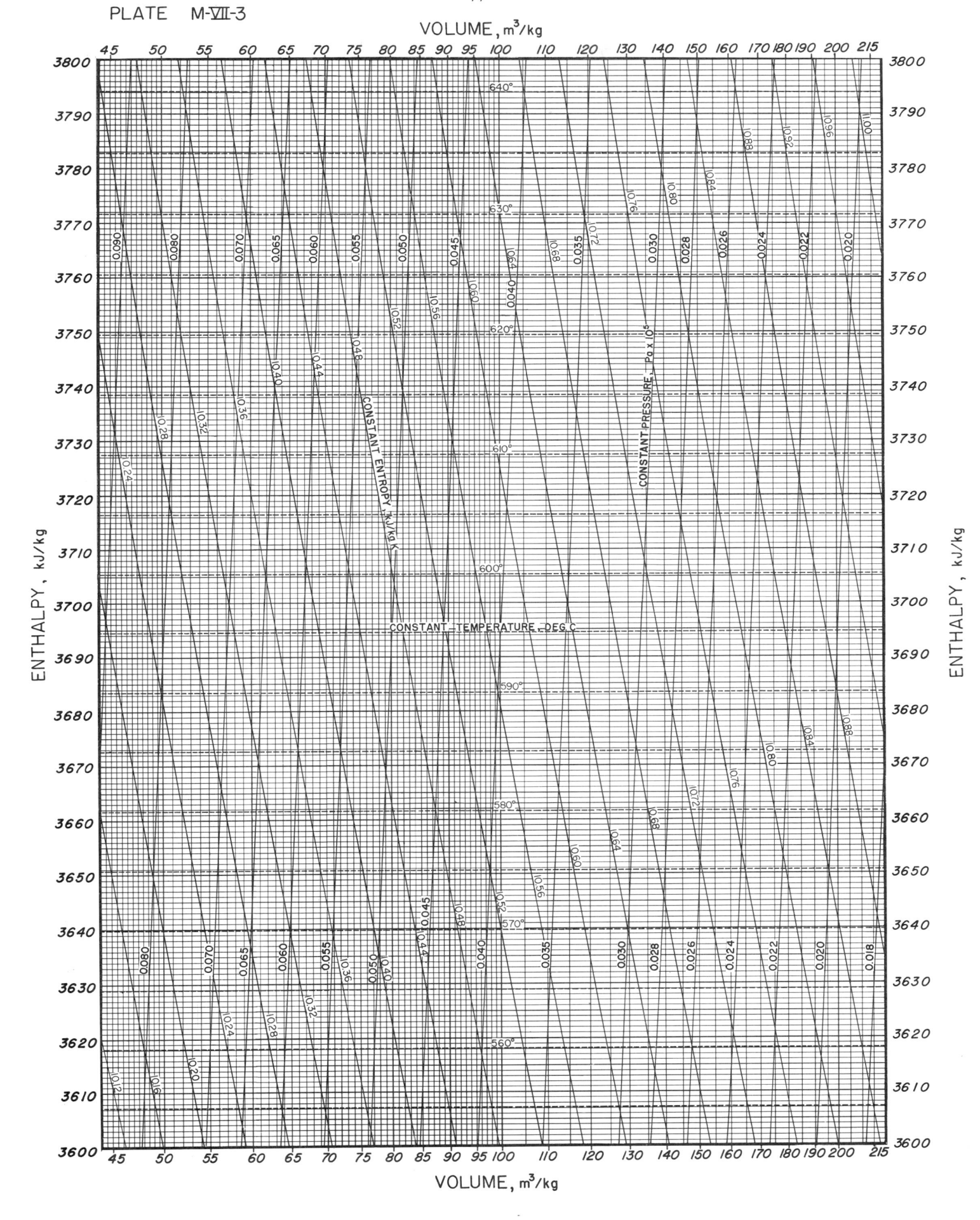

PLATE M-VII-4

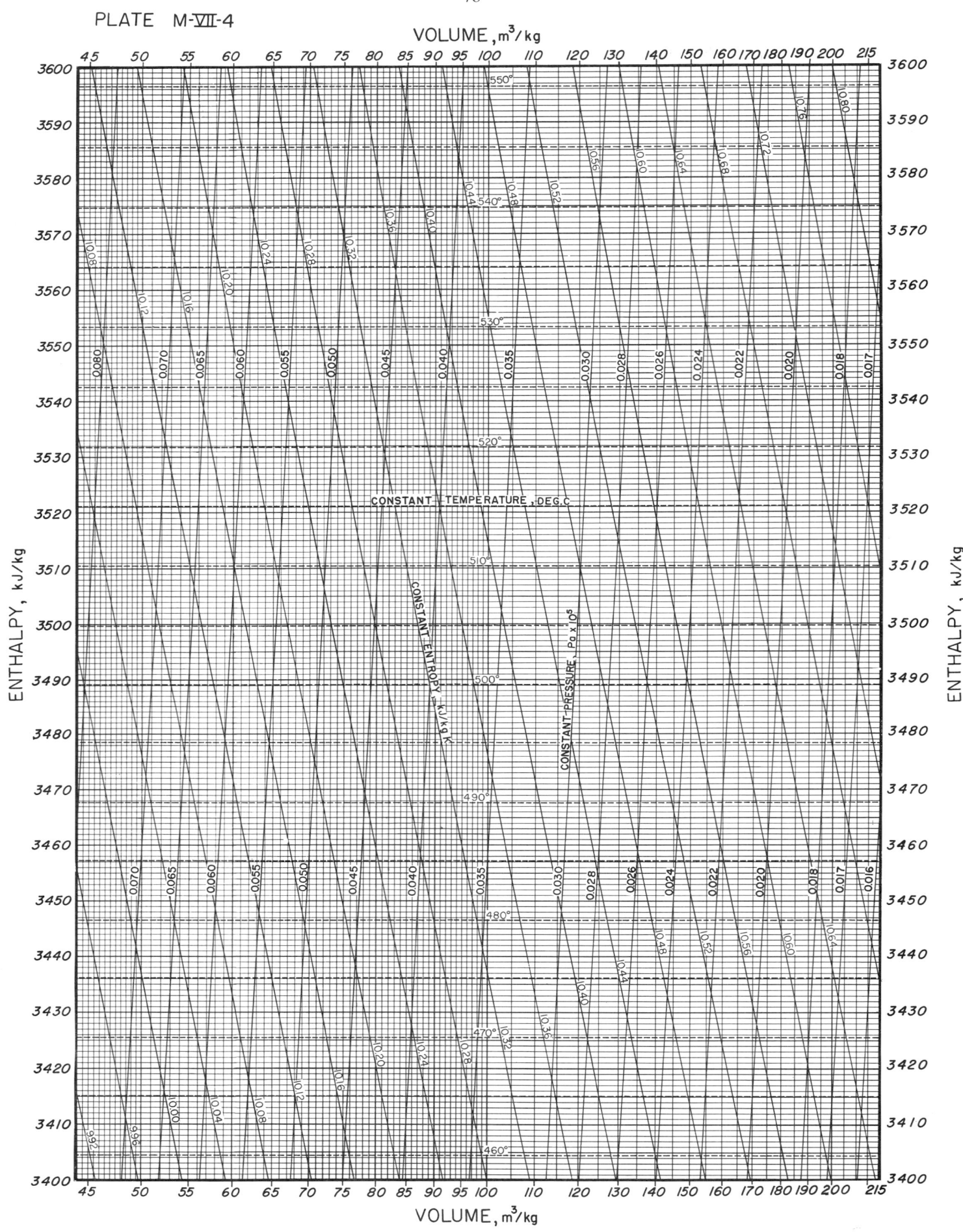

PLATE M-VII-5

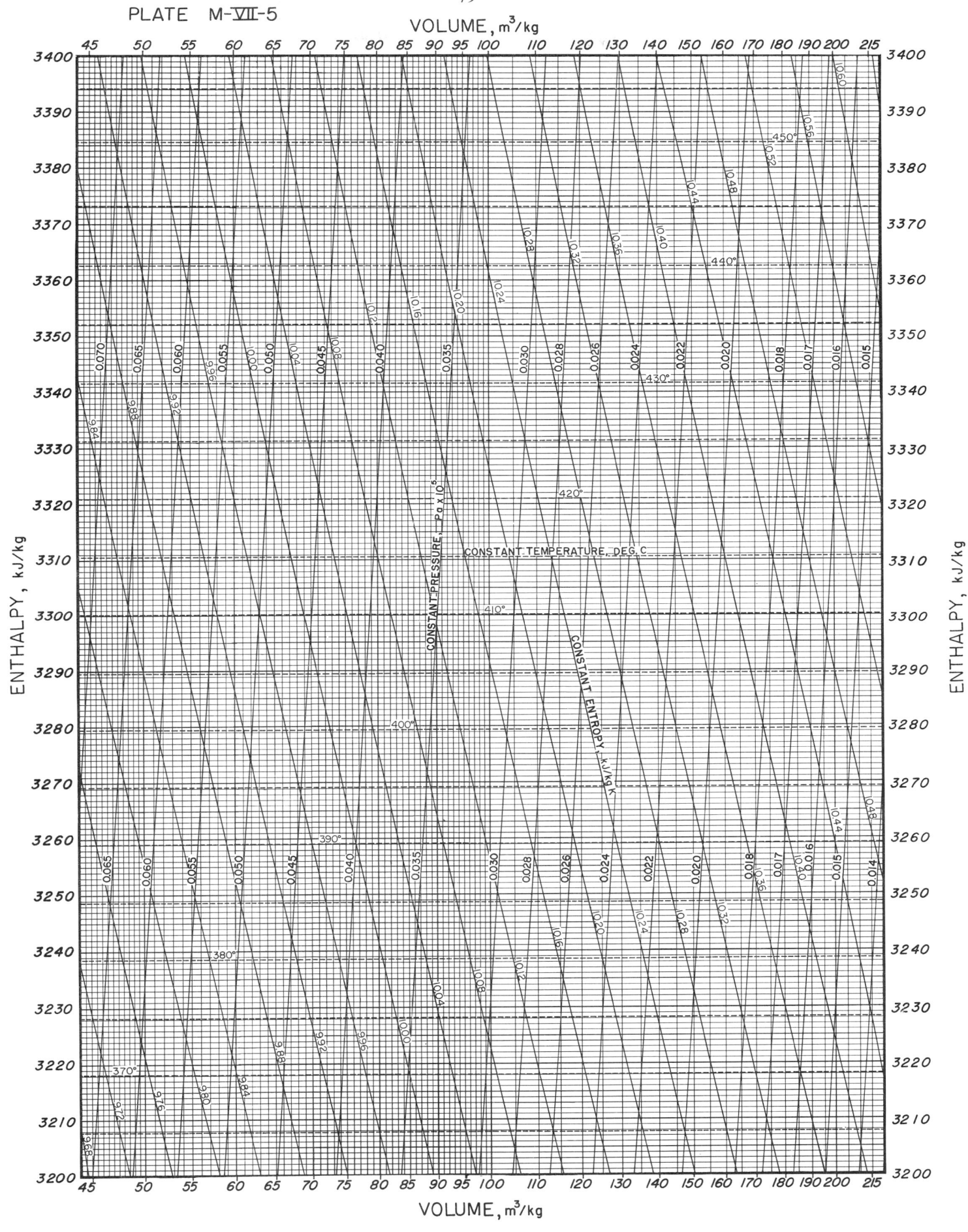

PLATE M-VII-6

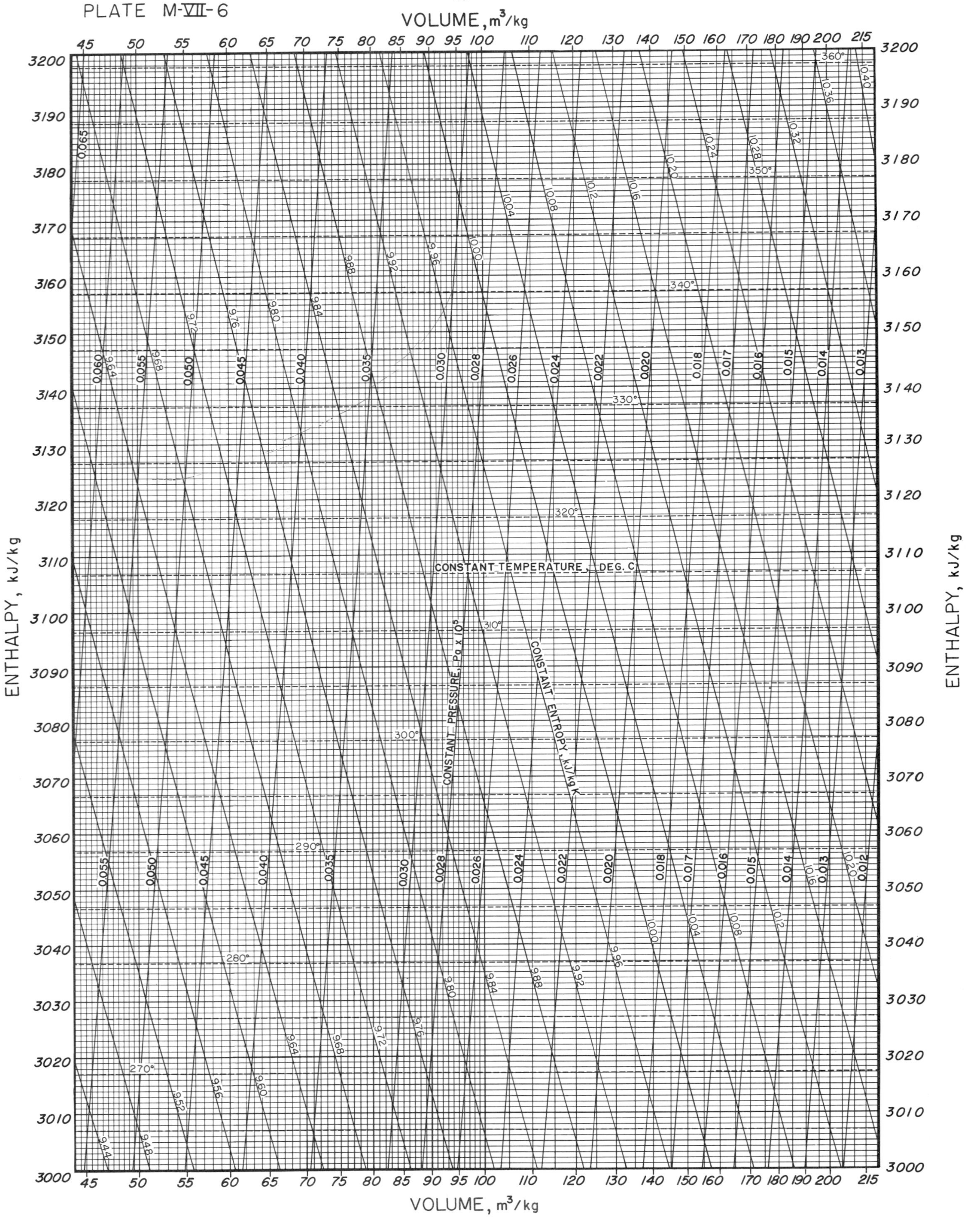
VOLUME, m³/kg
45 50 55 60 65 70 75 80 85 90 95 100 110 120 130 140 150 160 170 180 190 200 215
ENTHALPY, kJ/kg
3000 3010 3020 3030 3040 3050 3060 3070 3080 3090 3100 3110 3120 3130 3140 3150 3160 3170 3180 3190 3200
CONSTANT-TEMPERATURE, DEG. C
270° 280° 290° 300° 310° 320° 330° 340° 350° 360°
CONSTANT PRESSURE, Pa x 10⁵
0.065 0.060 0.055 0.050 0.045 0.040 0.035 0.030 0.028 0.026 0.024 0.022 0.020 0.018 0.017 0.016 0.015 0.014 0.013 0.012
CONSTANT ENTROPY, kJ/kg K
9.44 9.48 9.52 9.56 9.60 9.64 9.68 9.72 9.76 9.80 9.84 9.88 9.92 9.96 10.00 10.04 10.08 10.12 10.16 10.20 10.24 10.28 10.32 10.36 10.40
VOLUME, m³/kg

PLATE M-VII-7

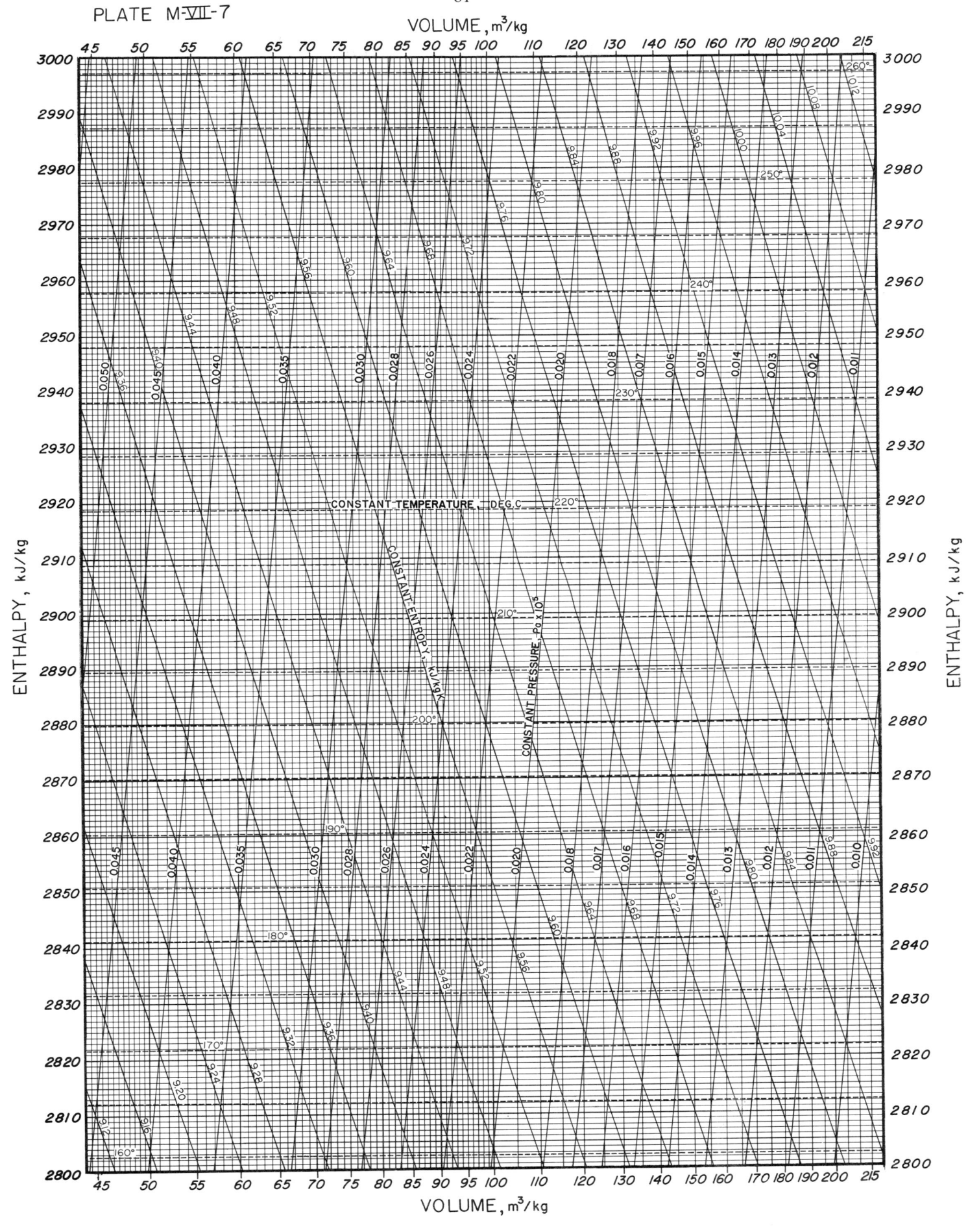

PLATE M-VII-8

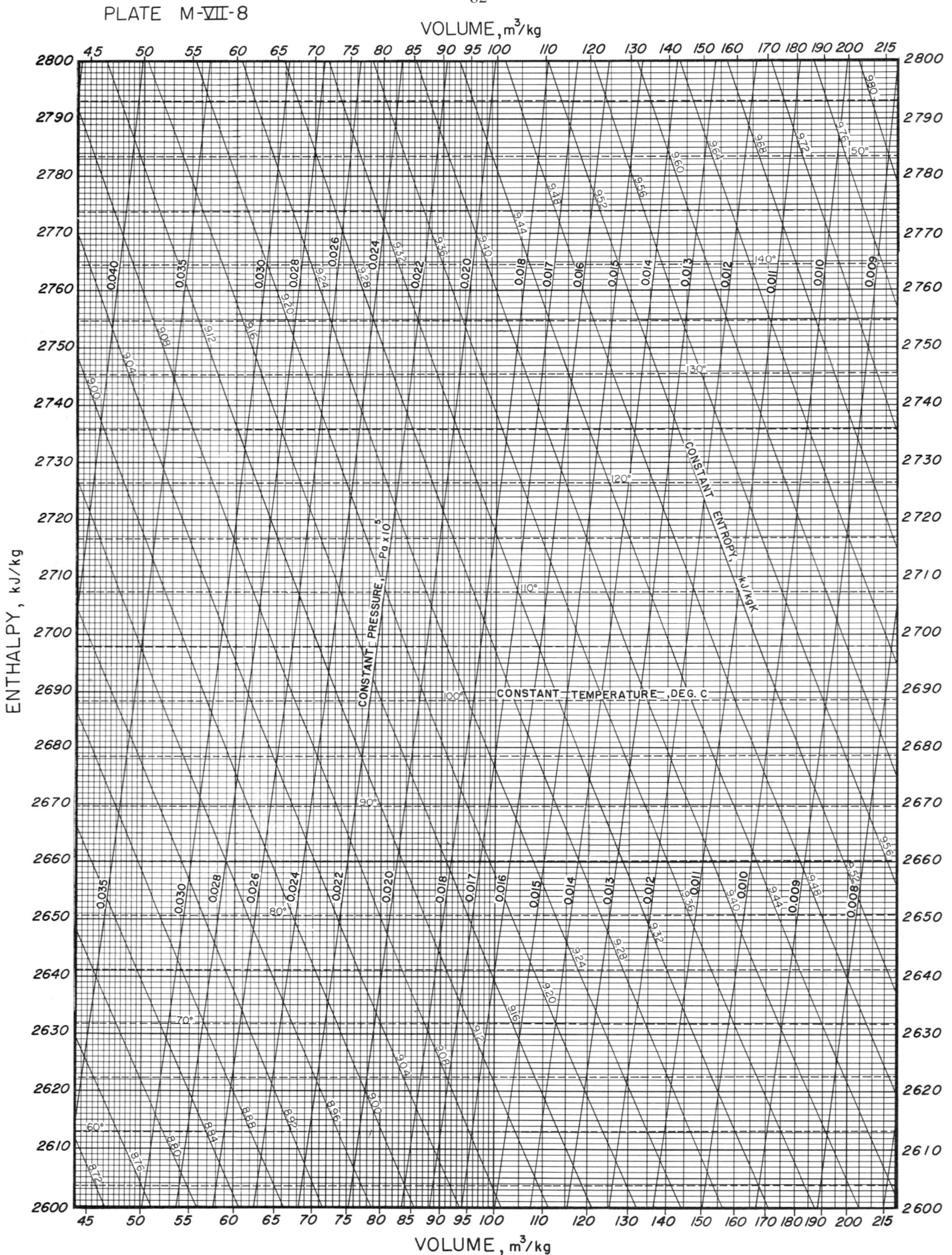

PLATE M-VII-9

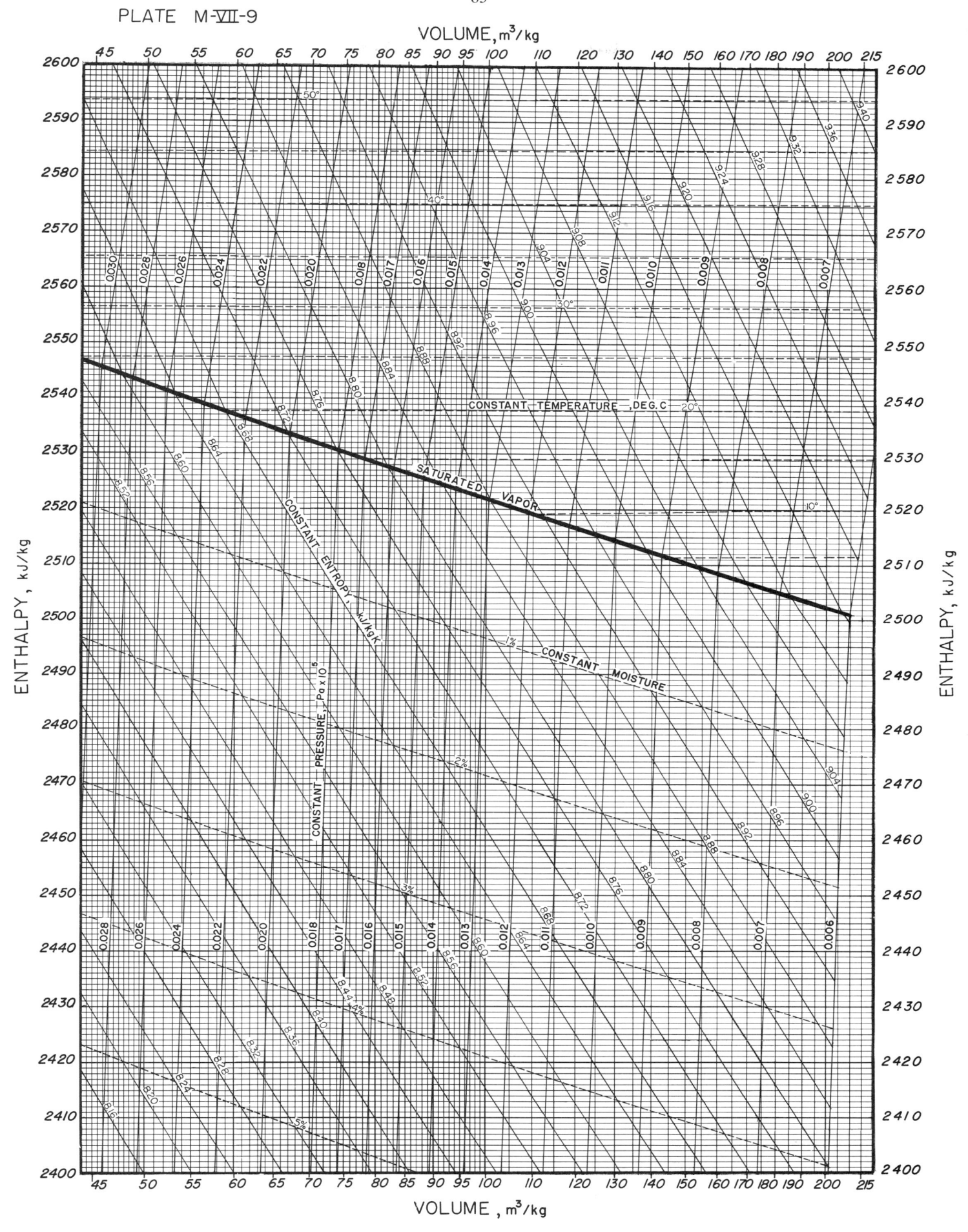

PLATE M-VII-10

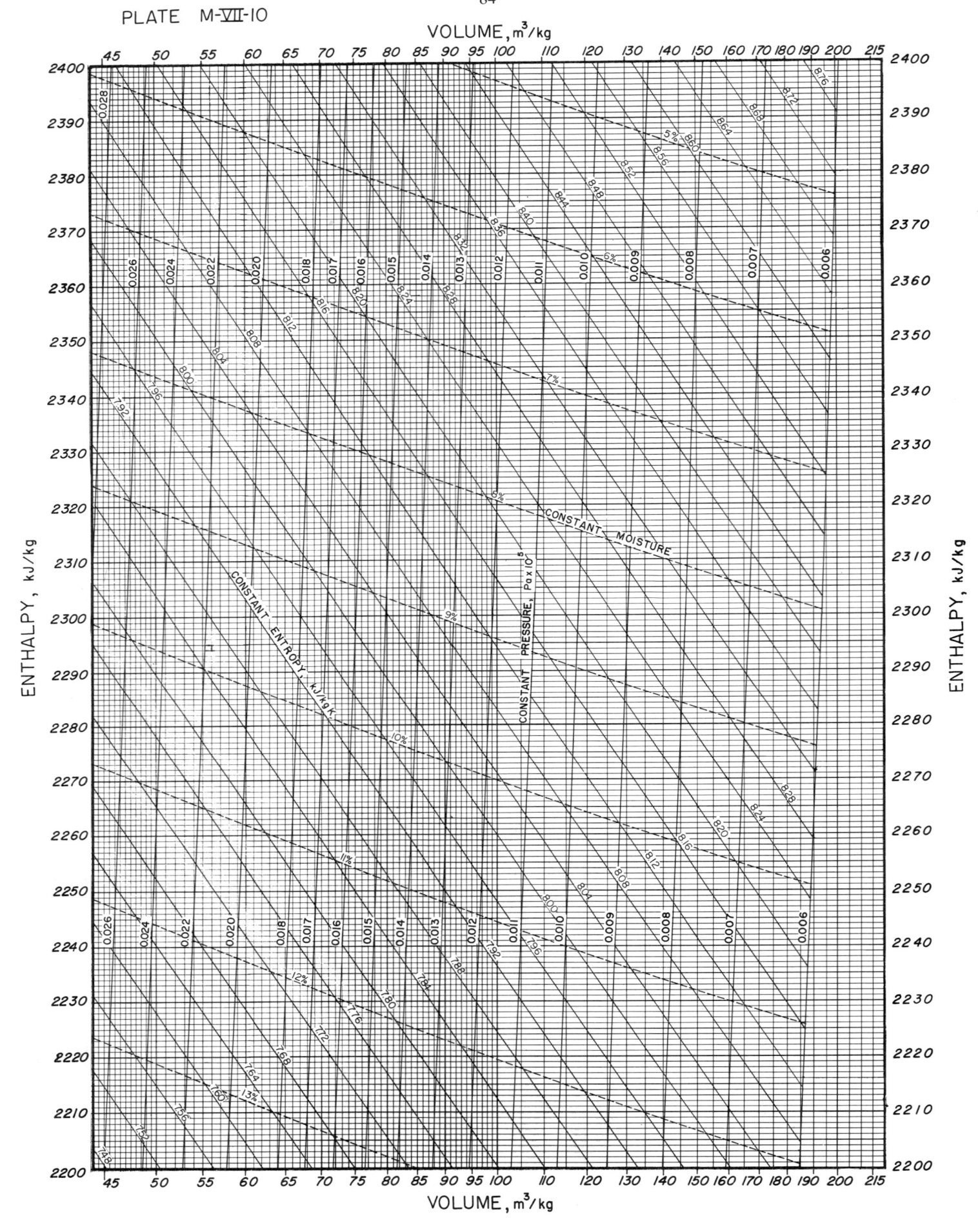

PLATE M-VII-11

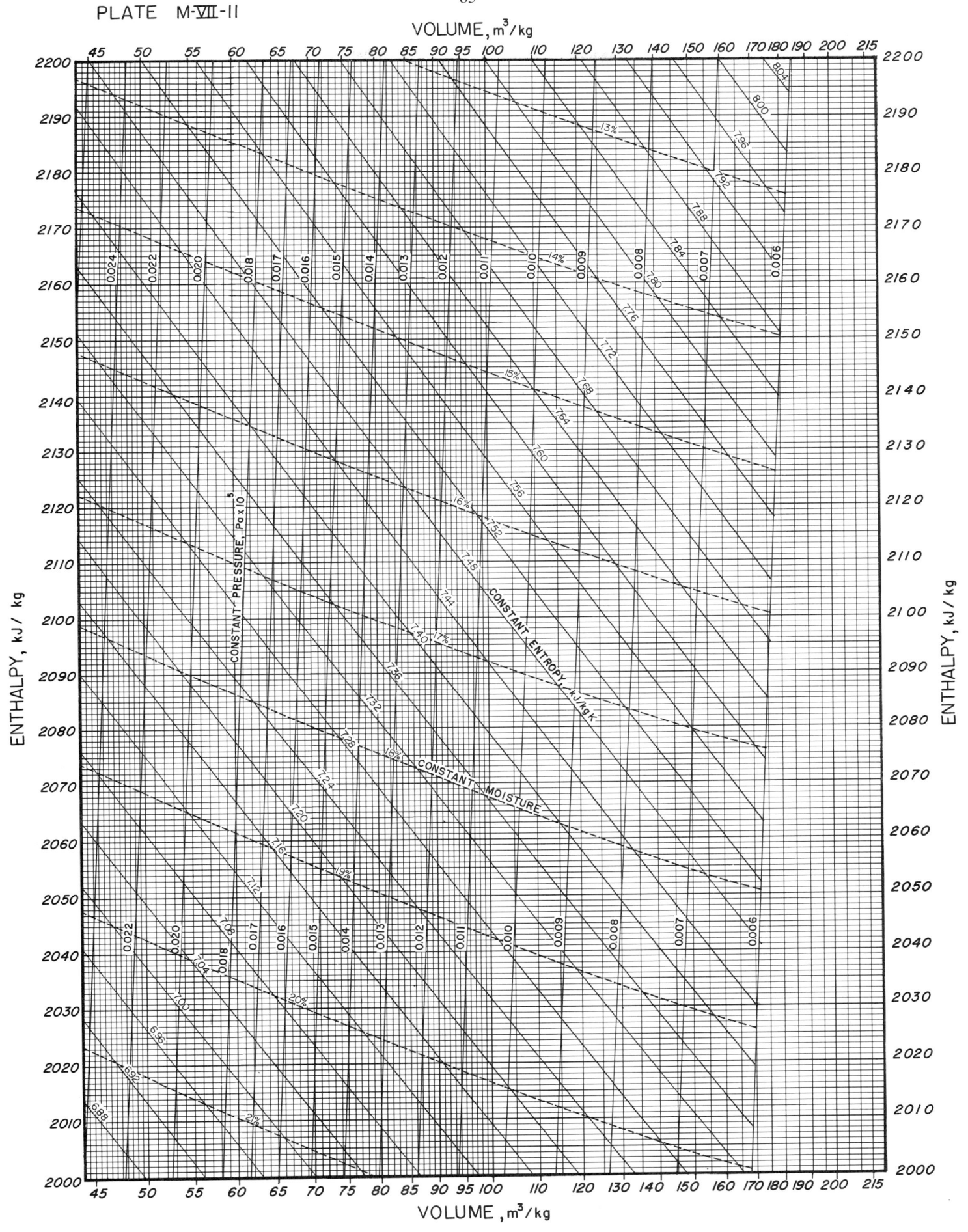

PLATE M-VII-12

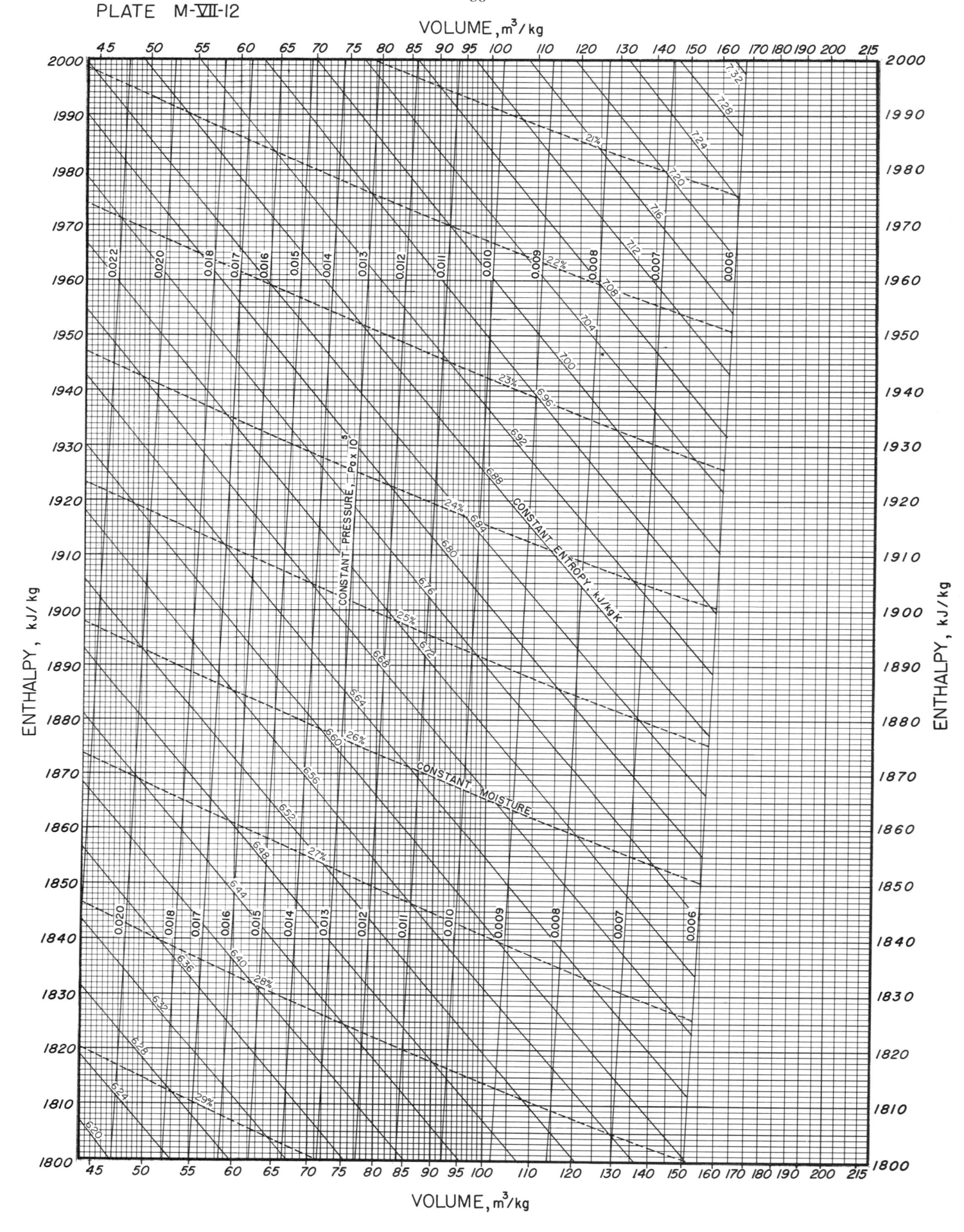

PLATE M-VIII

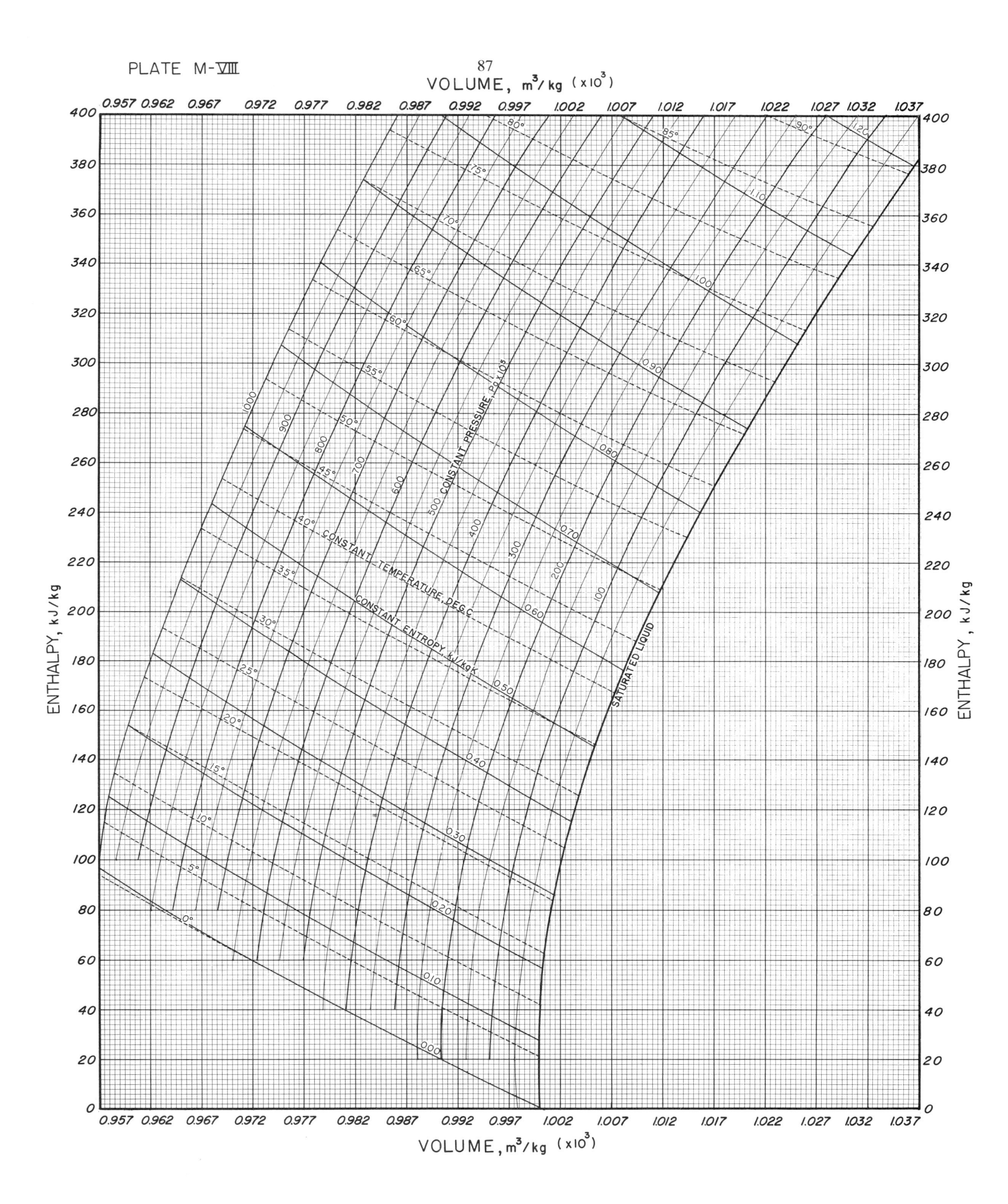

PLATE M-IX

PLATE M-X

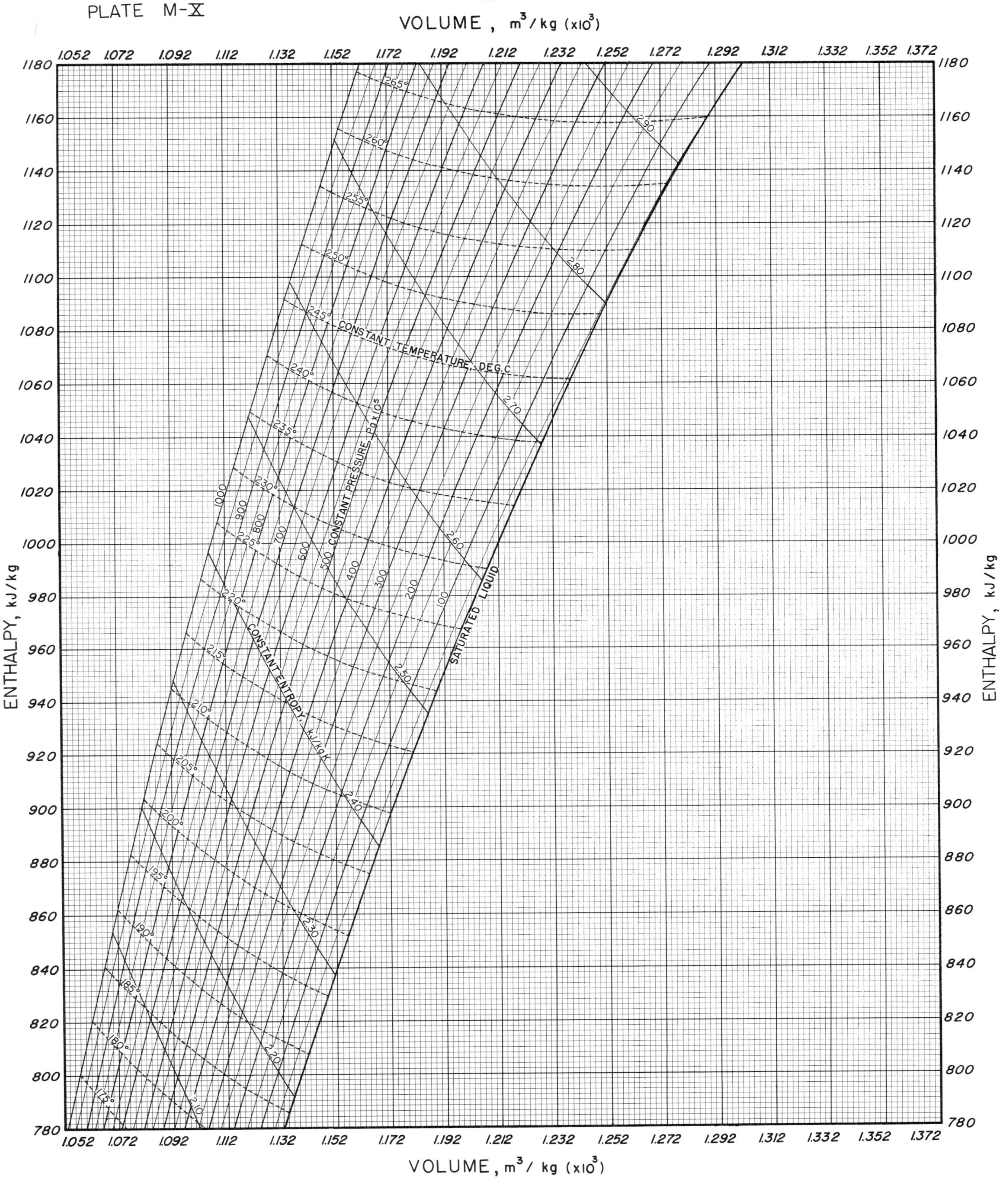
VOLUME, m³/kg (x10³)
1.052
1.072
1.092
1.112
1.132
1.152
1.172
1.192
1.212
1.232
1.252
1.272
1.292
1.312
1.332
1.352
1.372
ENTHALPY, kJ/kg
1180
1160
1140
1120
1100
1080
1060
1040
1020
1000
980
960
940
920
900
880
860
840
820
800
780
CONSTANT TEMPERATURE, DEG.C
265°
260°
255°
250°
245°
240°
235°
230°
225°
220°
215°
210°
205°
200°
195°
190°
185°
180°
175°
CONSTANT PRESSURE, Pa x 10⁵
1000
900
800
700
600
500
400
300
200
100
CONSTANT ENTROPY, kJ/kg K
2.10
2.20
2.30
2.40
2.50
2.60
2.70
2.80
2.90
SATURATED LIQUID

PLATE M-XI

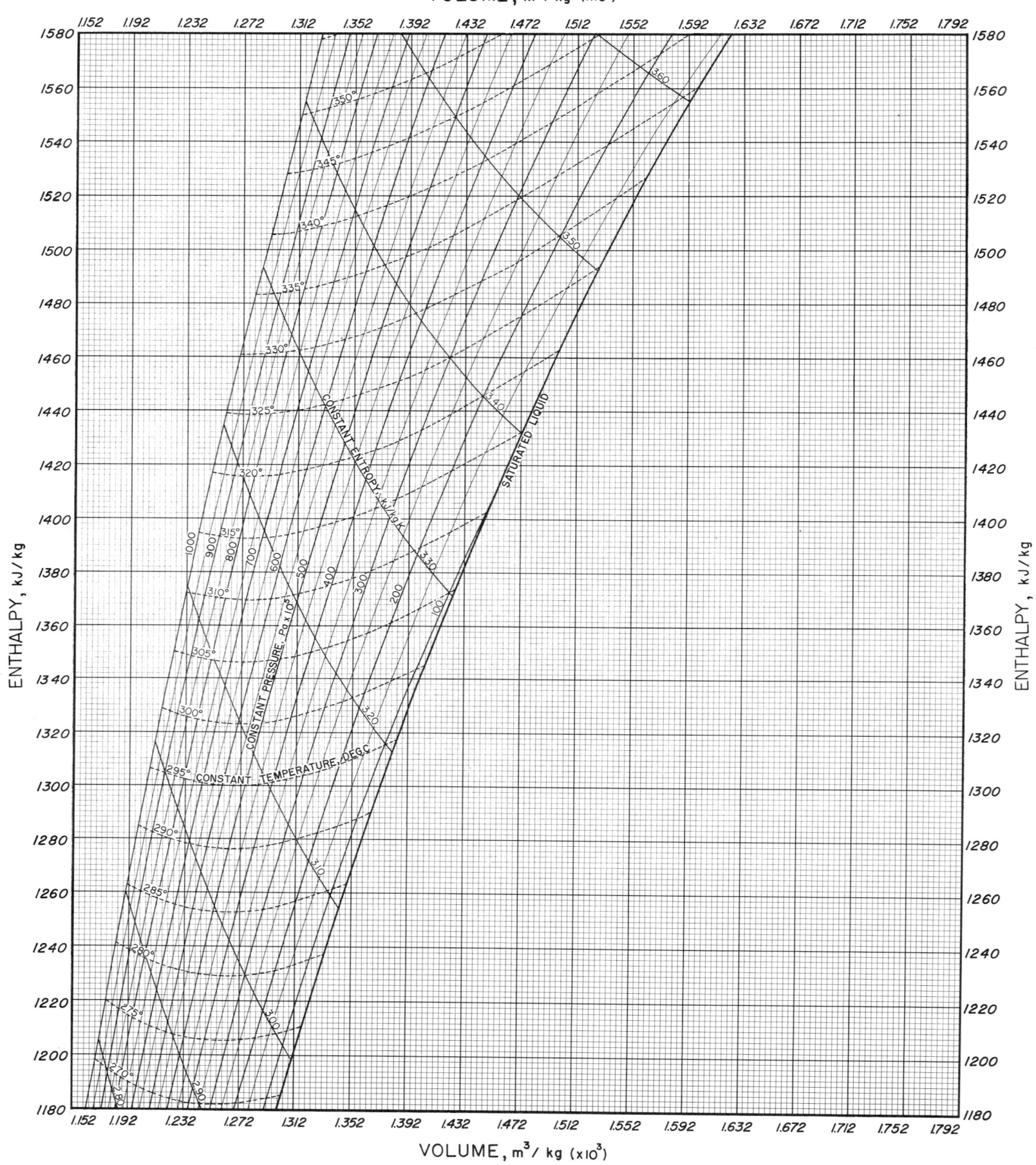

PLATE M-XII

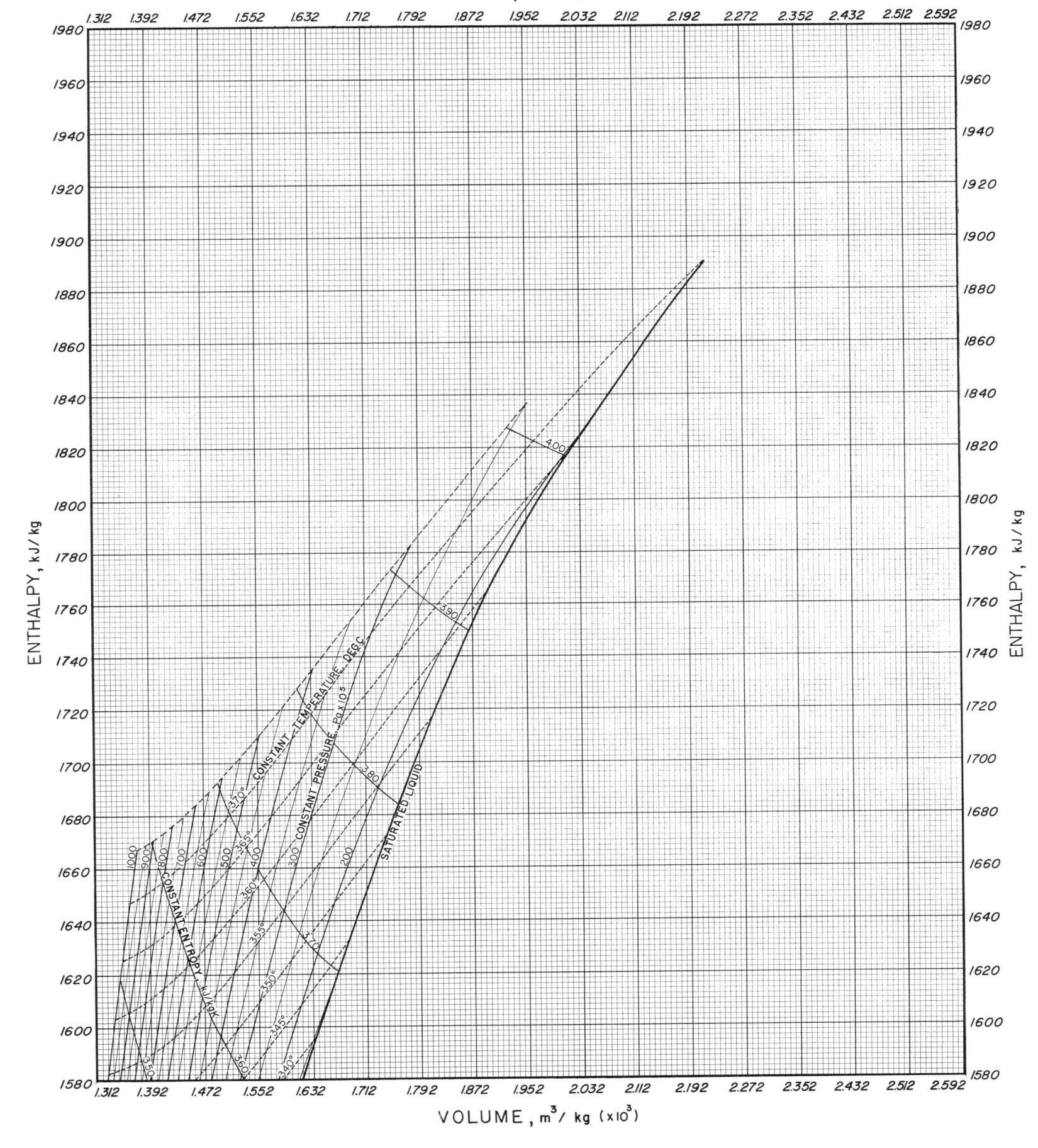

U.S. CUSTOMARY CHARTS

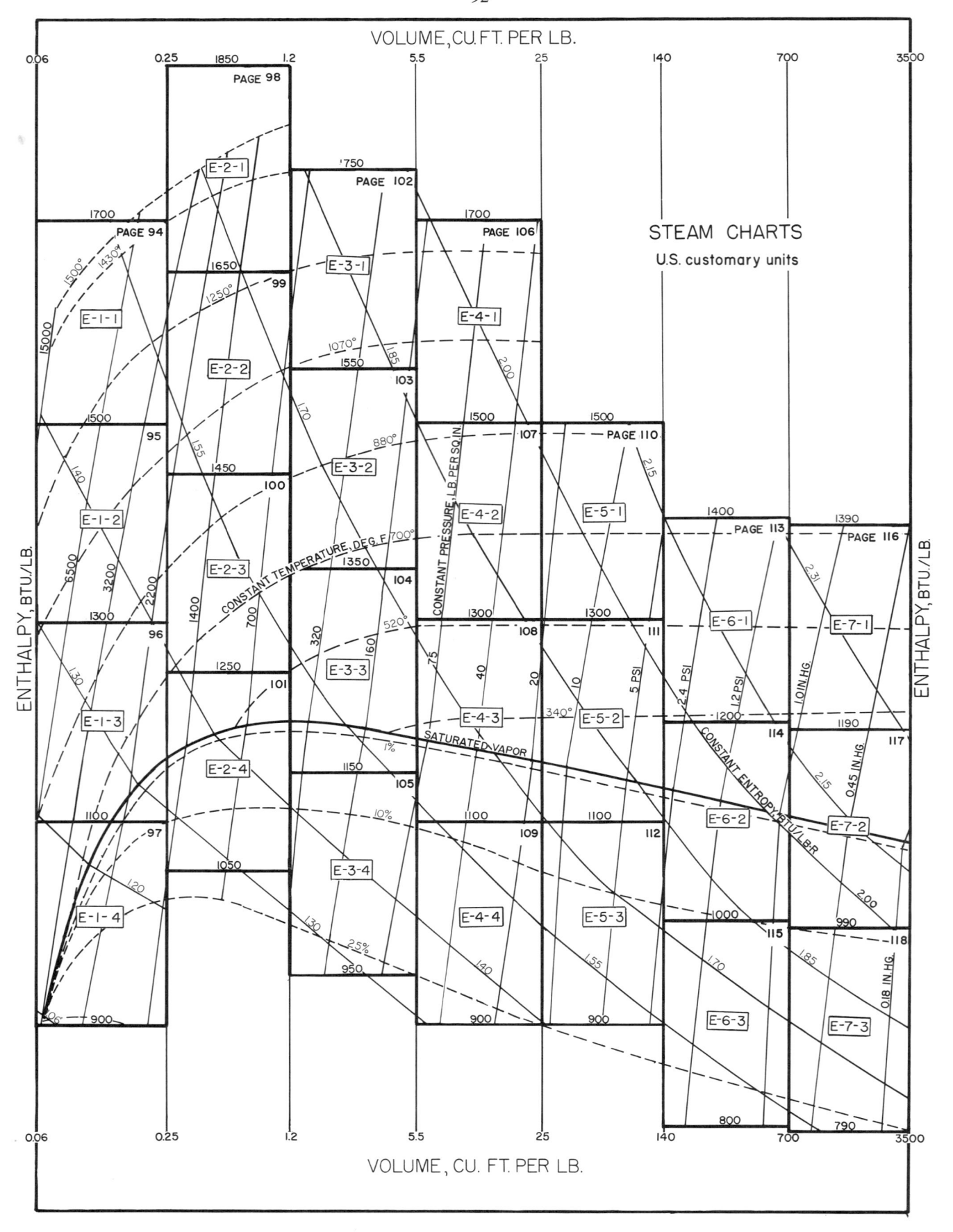

STEAM CHARTS
U.S. customary units

93

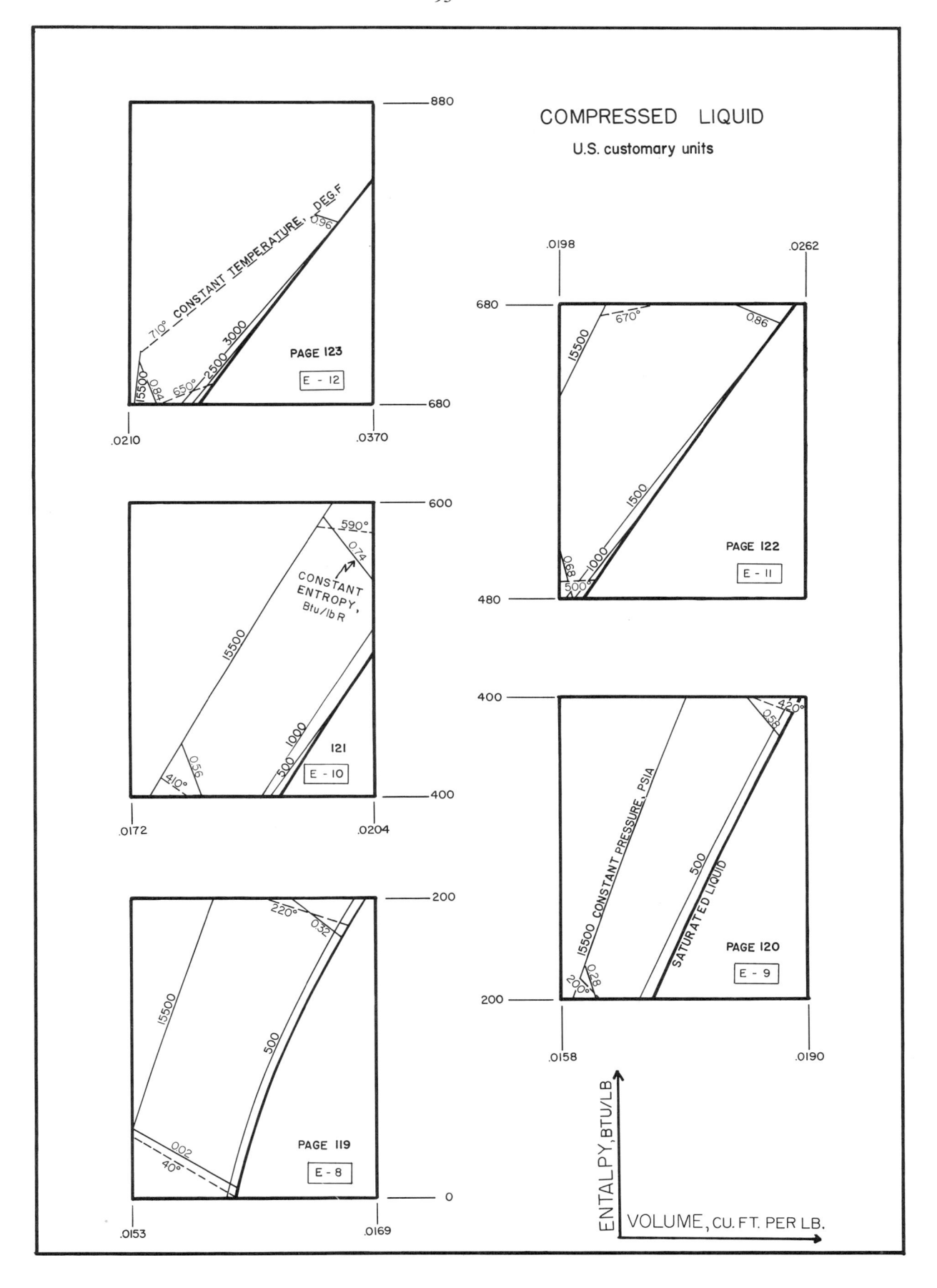

COMPRESSED LIQUID
U.S. customary units

PLATE E-I-I

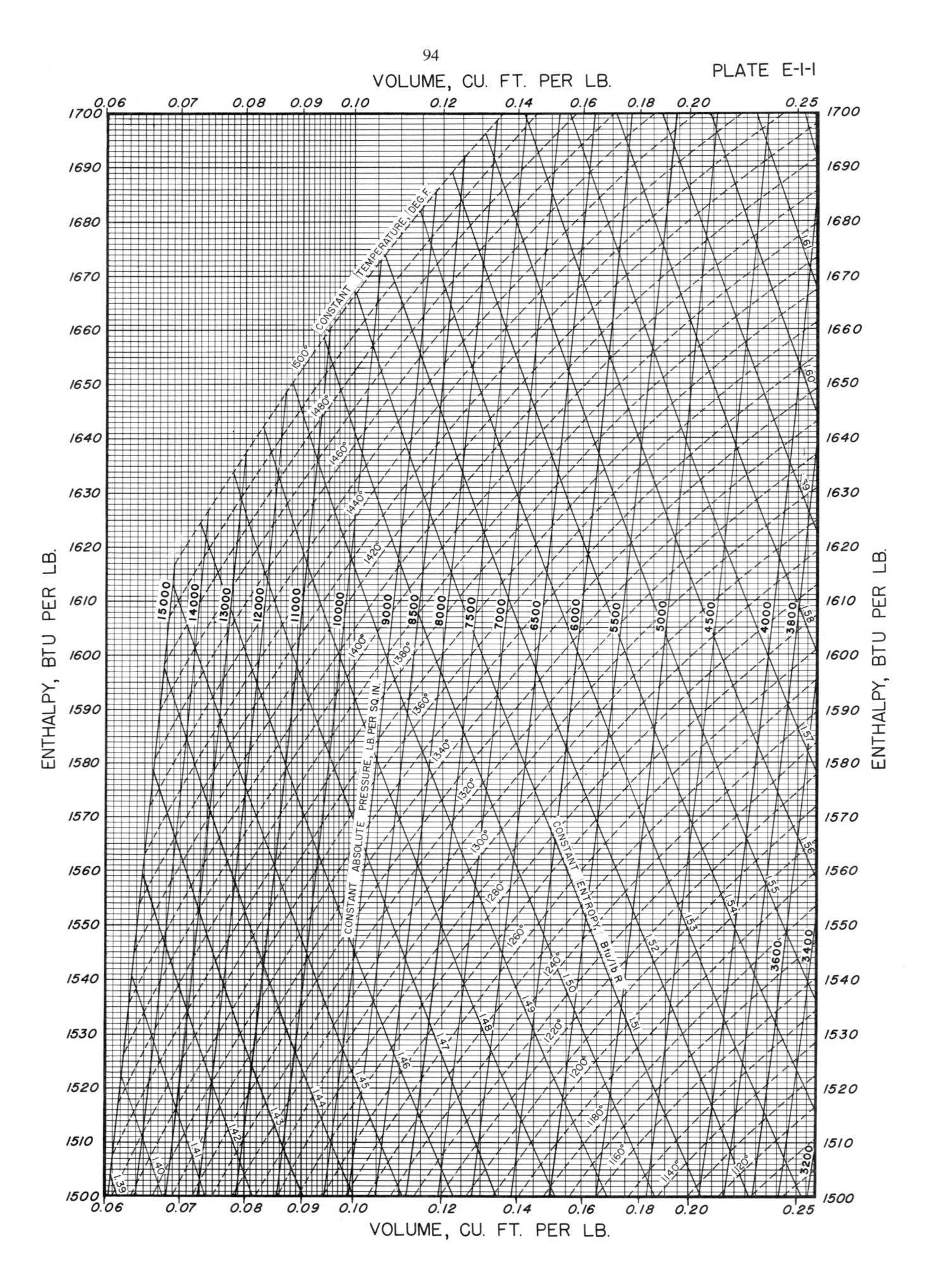

PLATE E-I-2

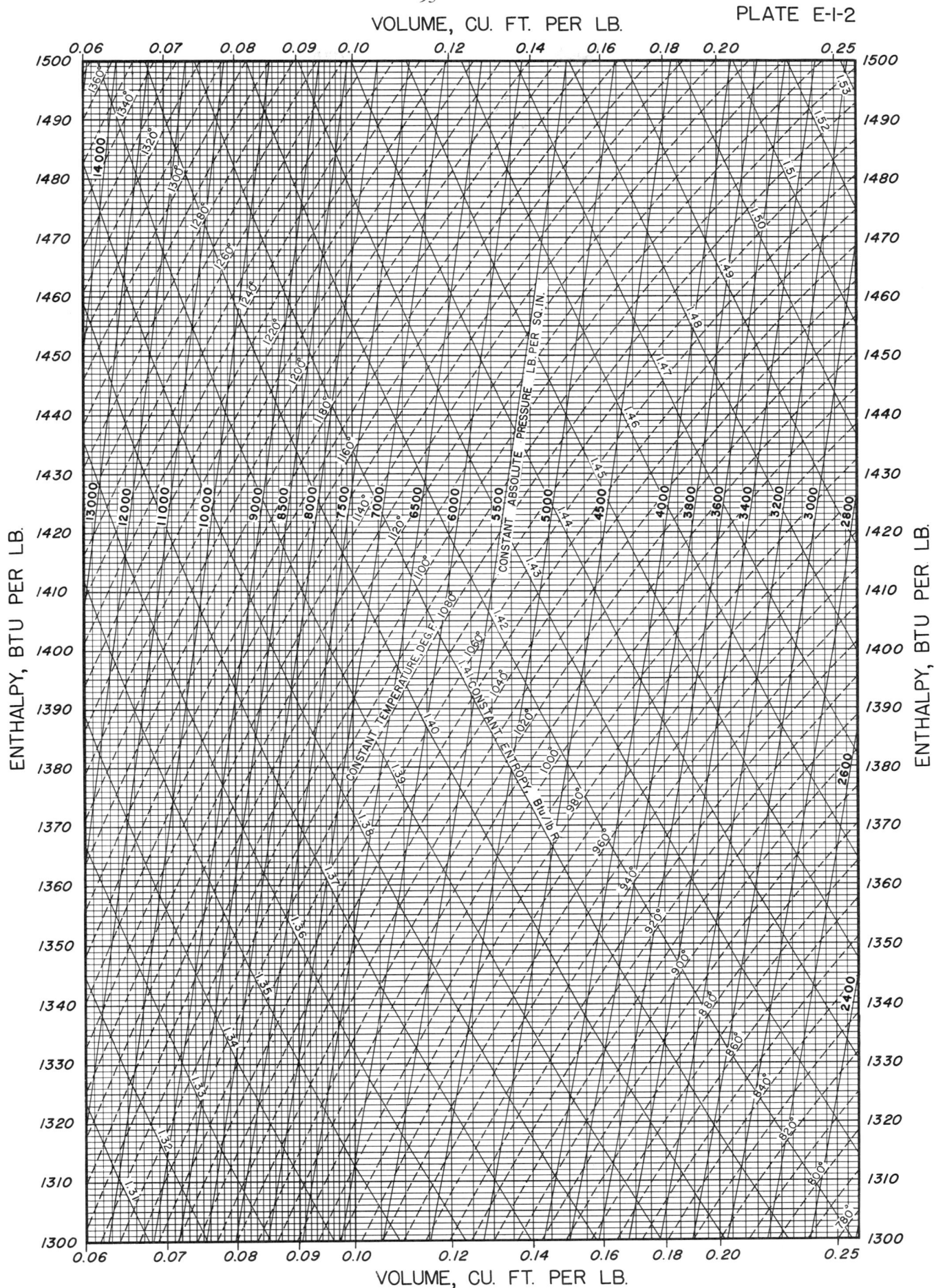

PLATE E-1-3

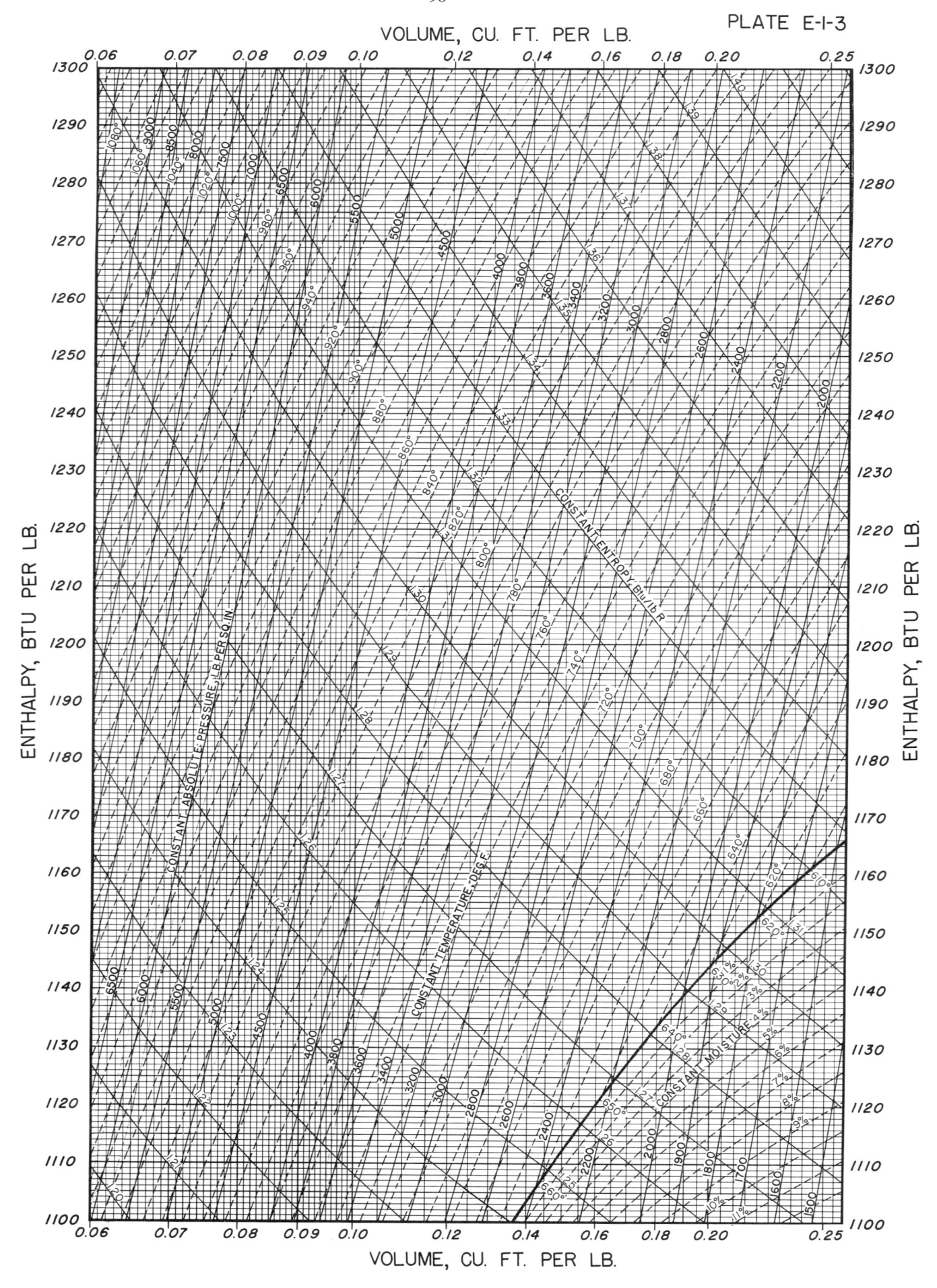

PLATE E-1-4

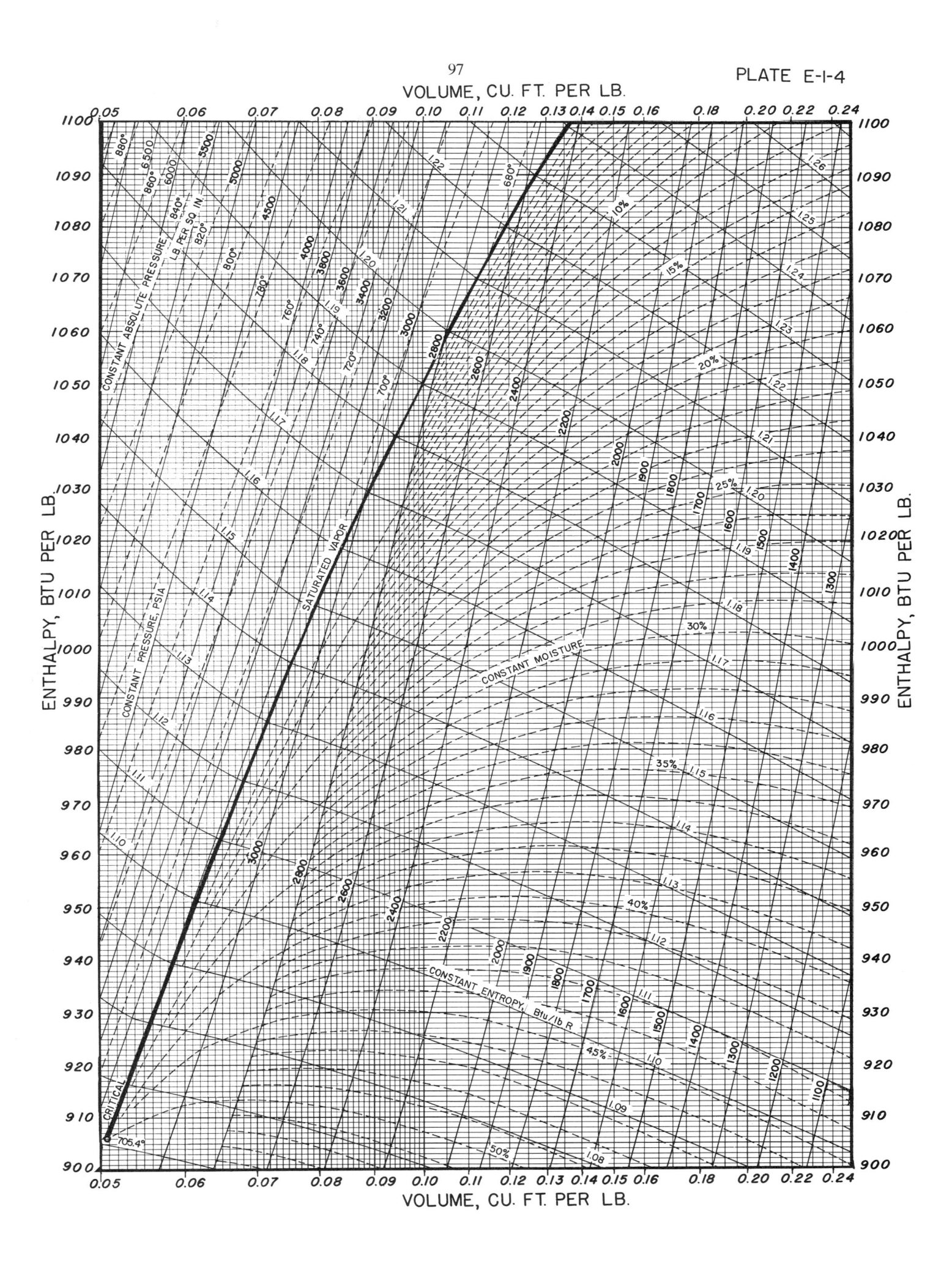

PLATE E-2-1

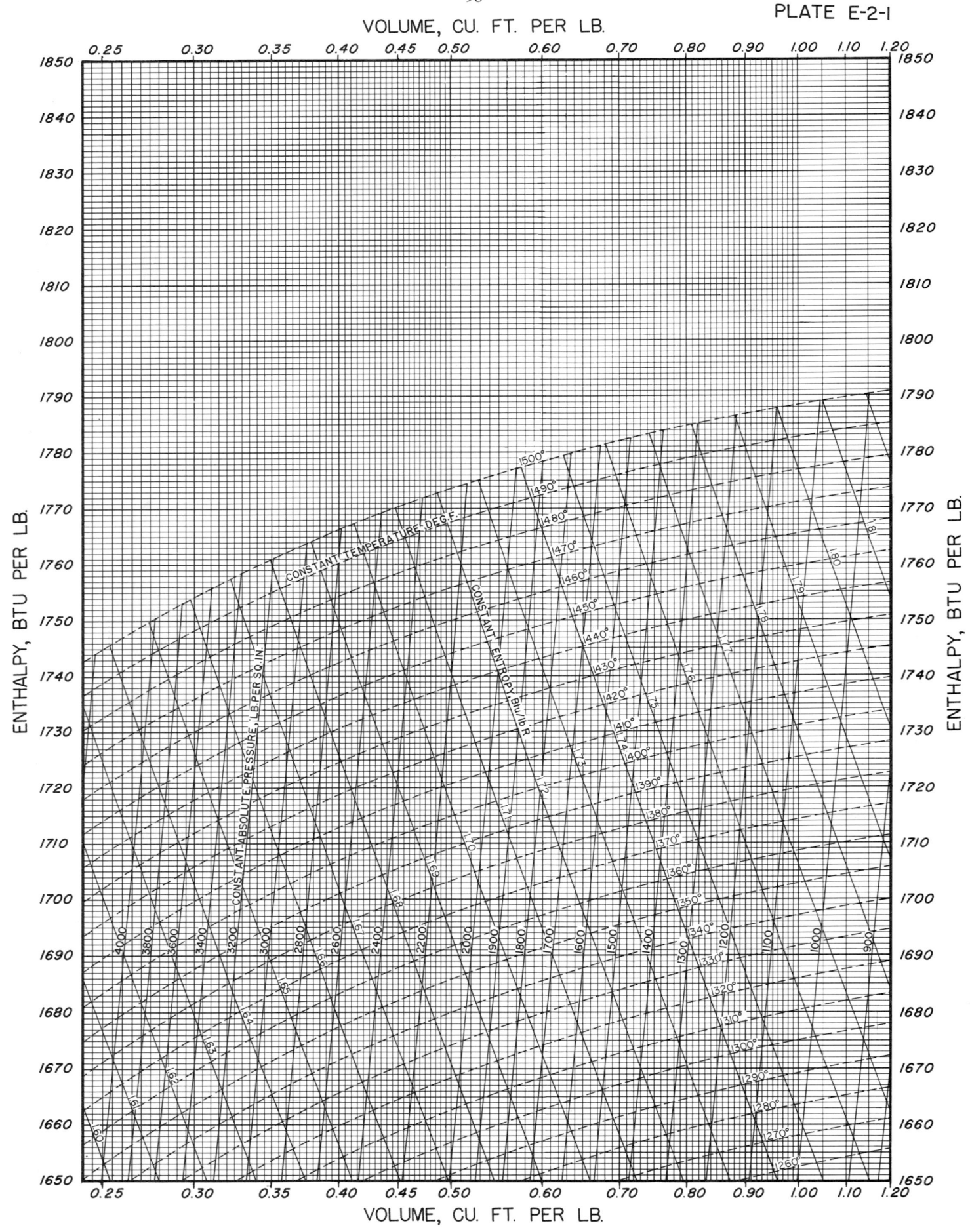

PLATE E-2-2

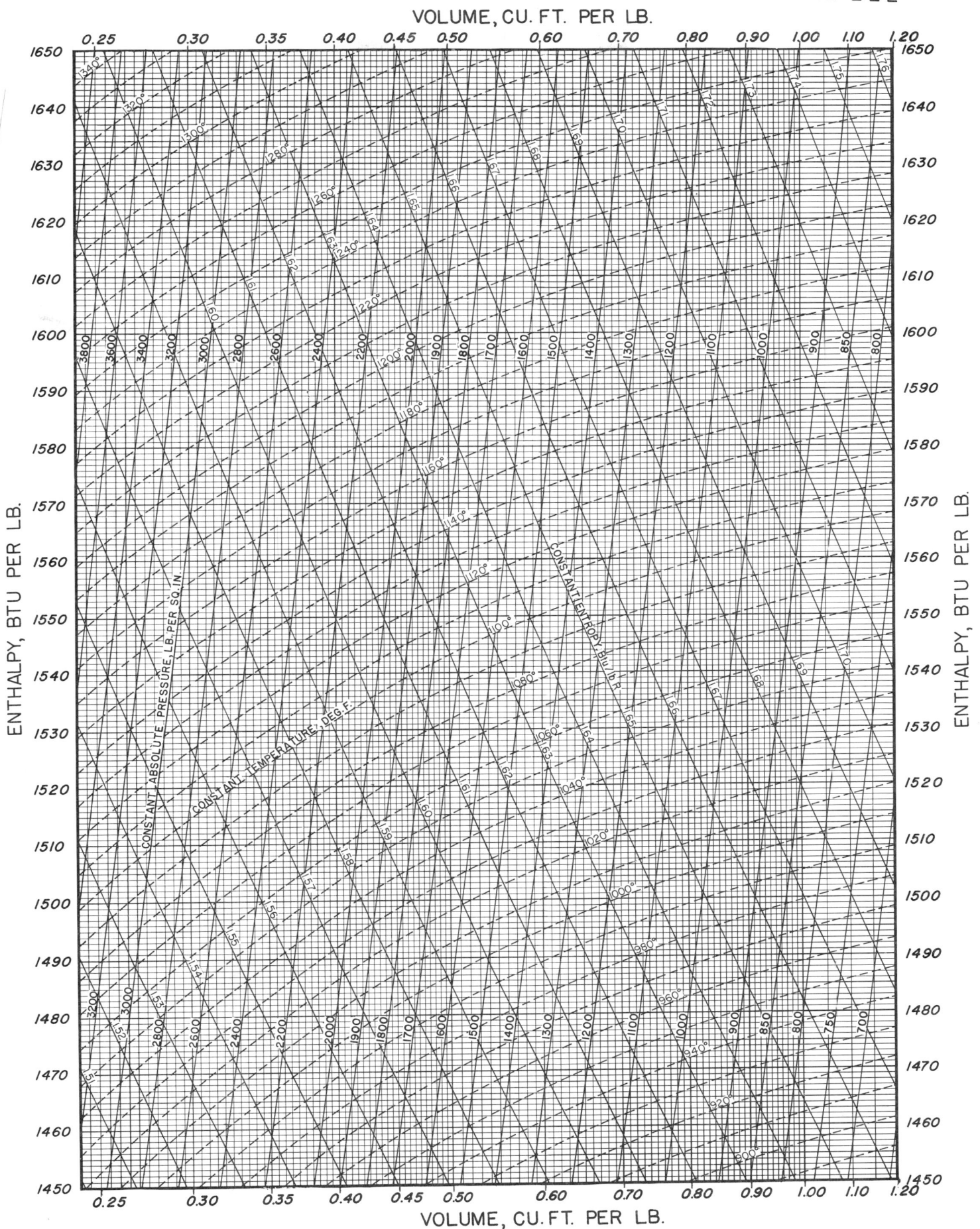

PLATE E-2-3

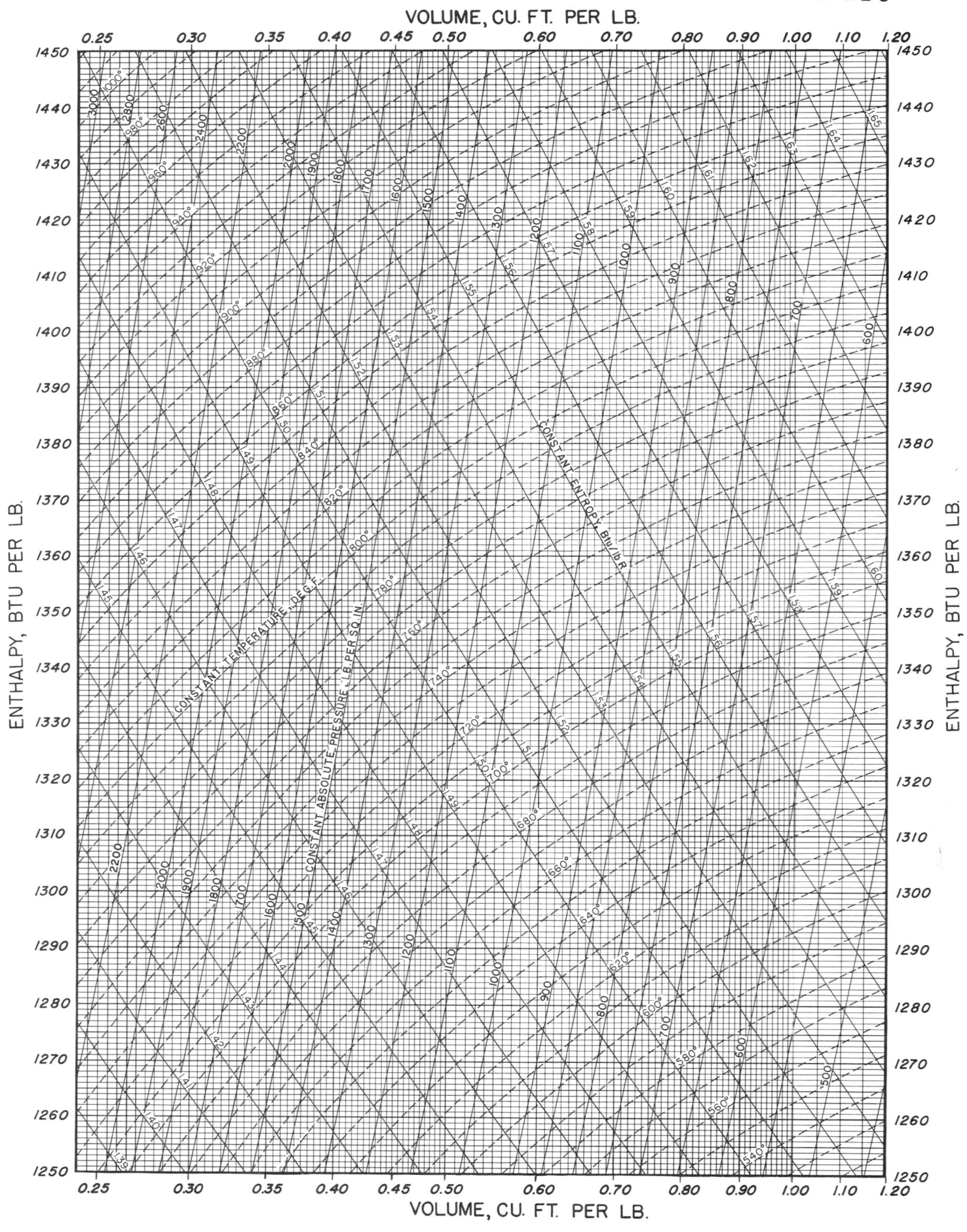

PLATE E-2-4

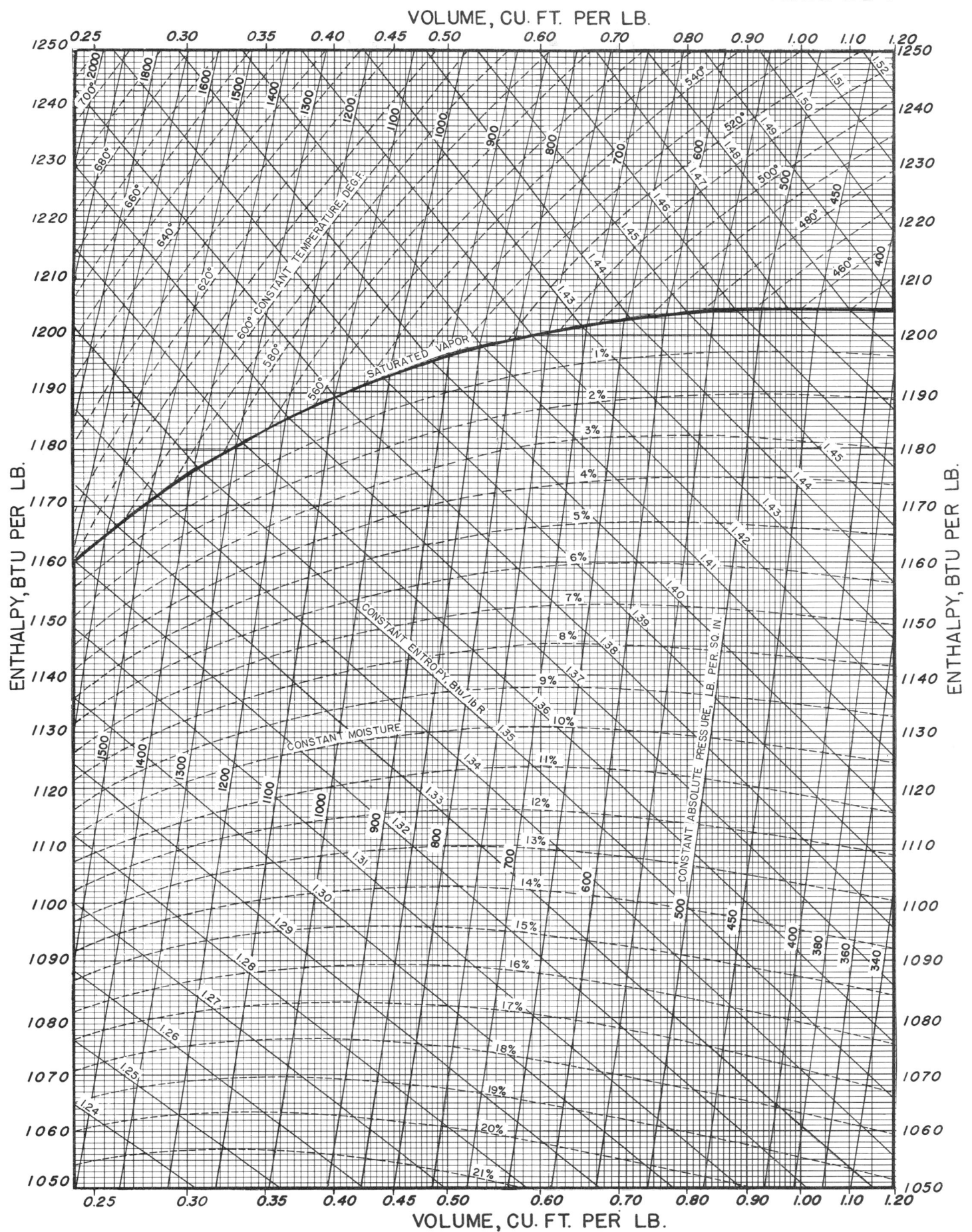

PLATE E-3-I

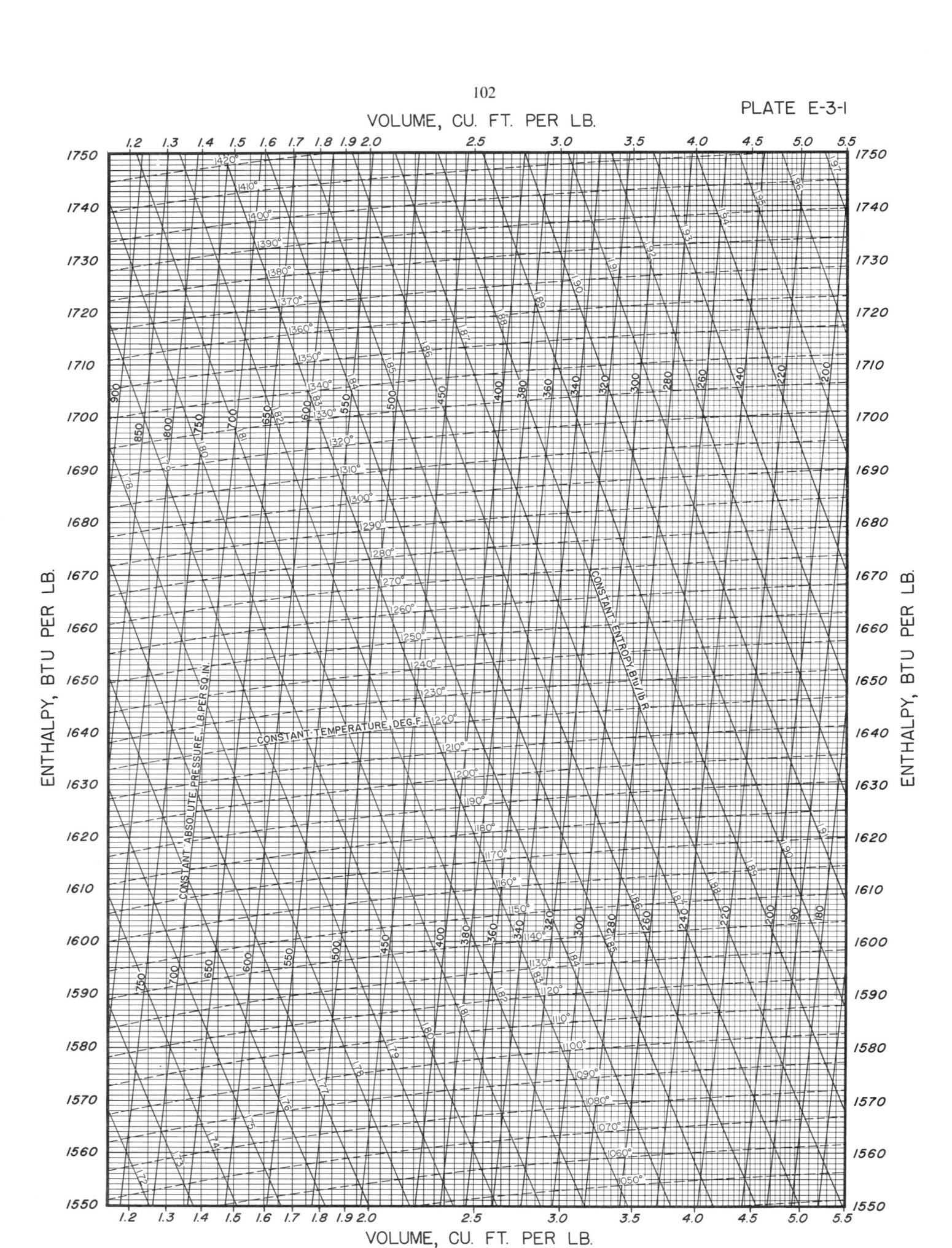

PLATE E-3-2

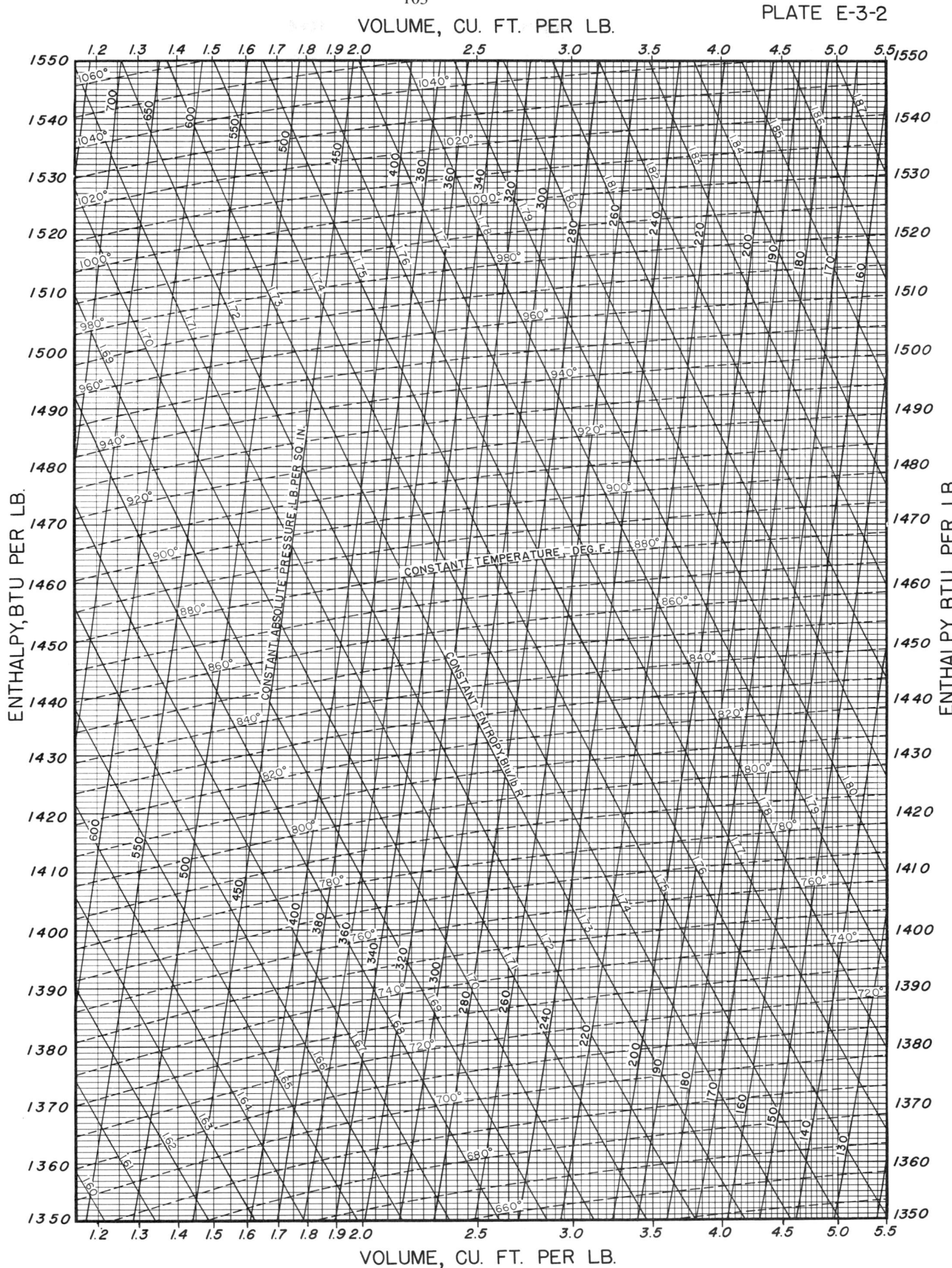

PLATE E-3-3

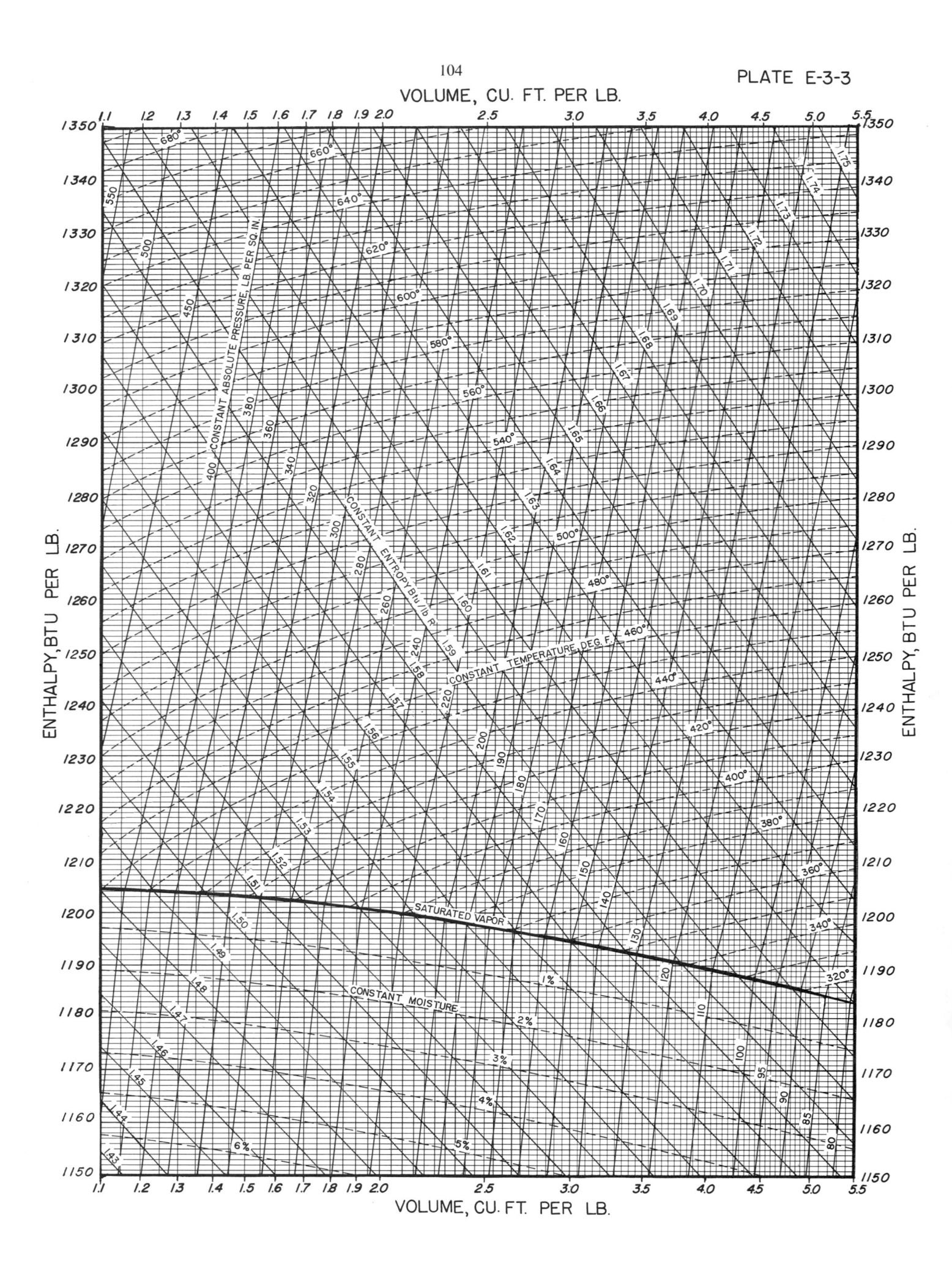

PLATE E-3-4

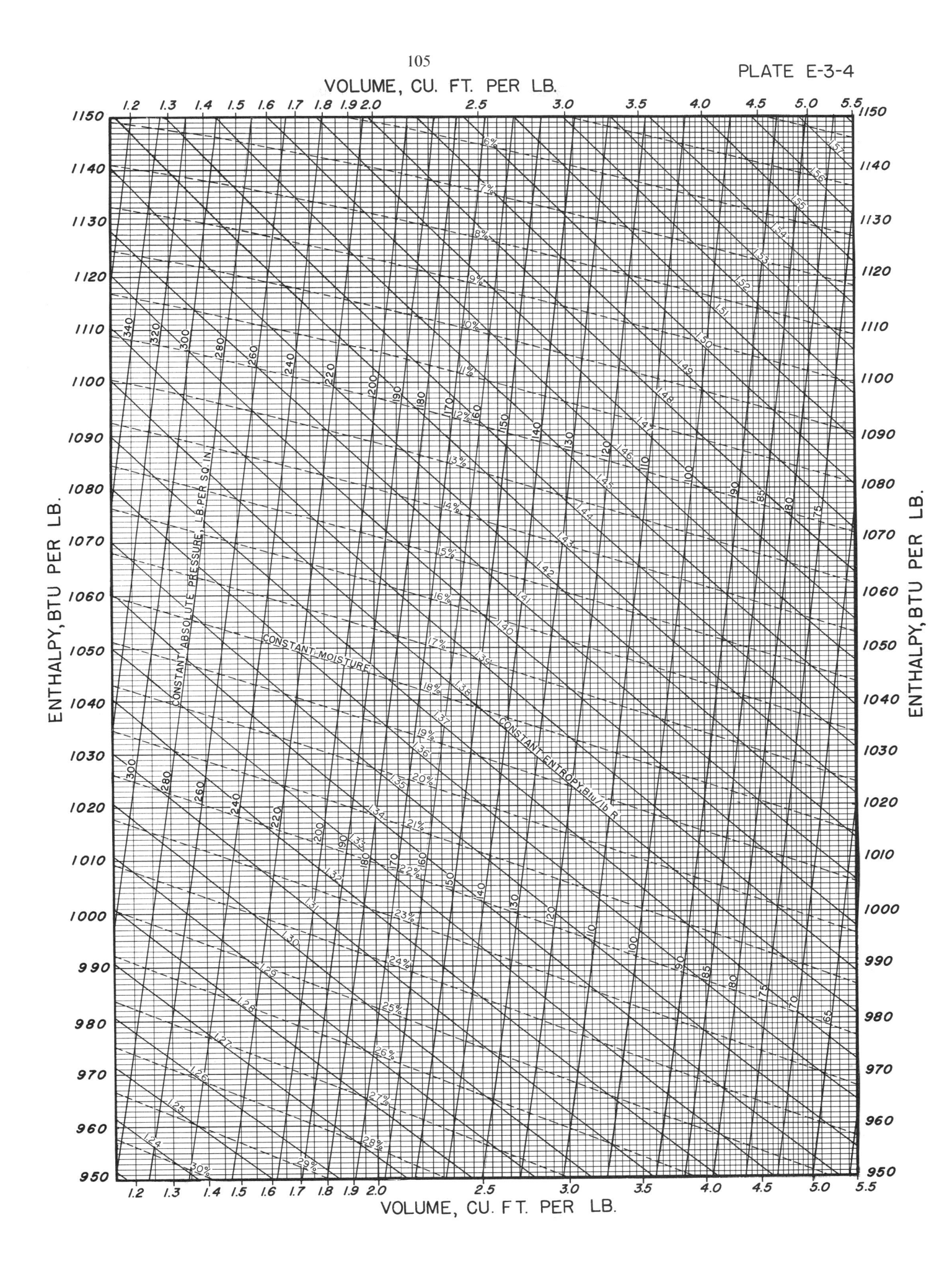

PLATE E-4-1

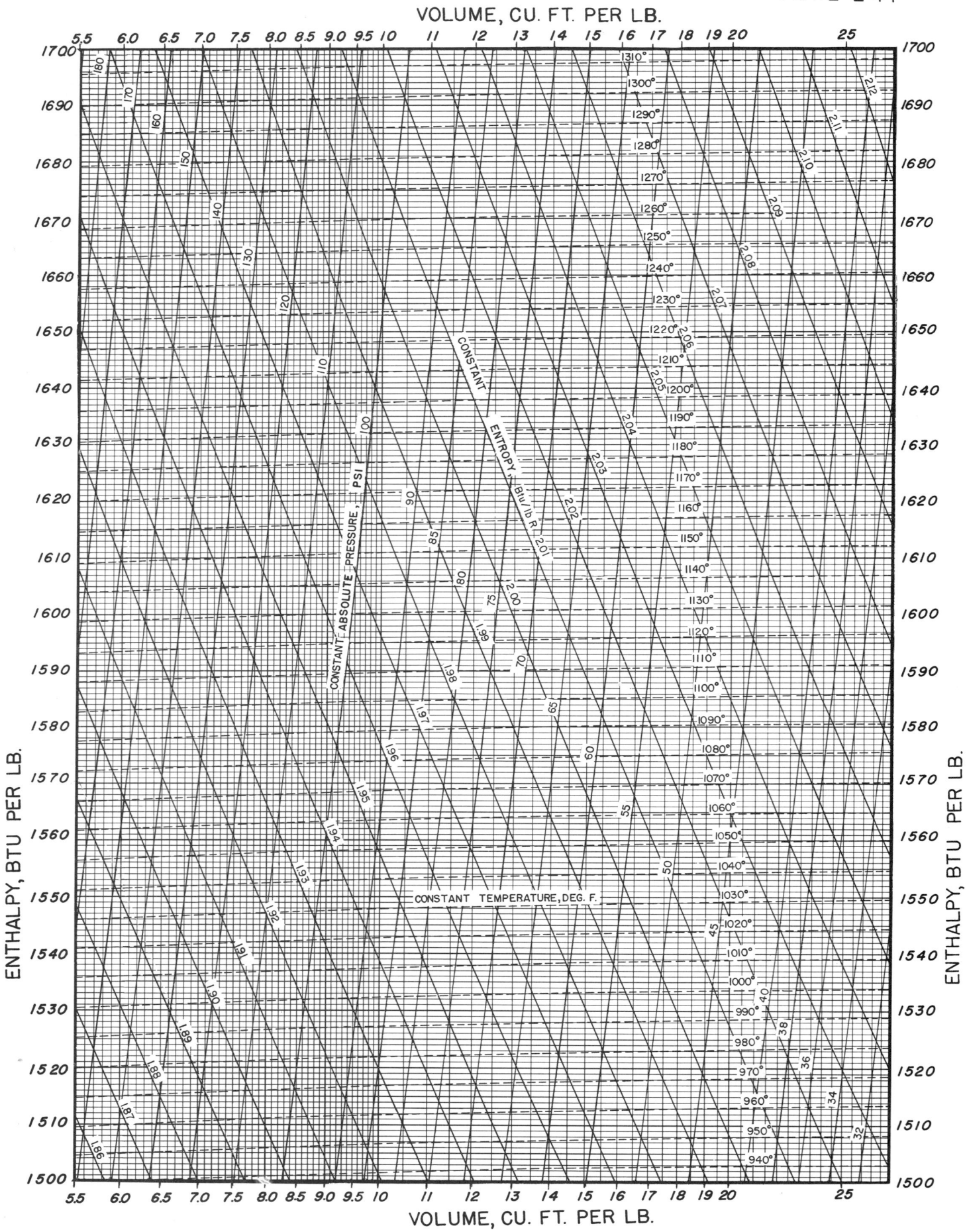

PLATE E-4-2

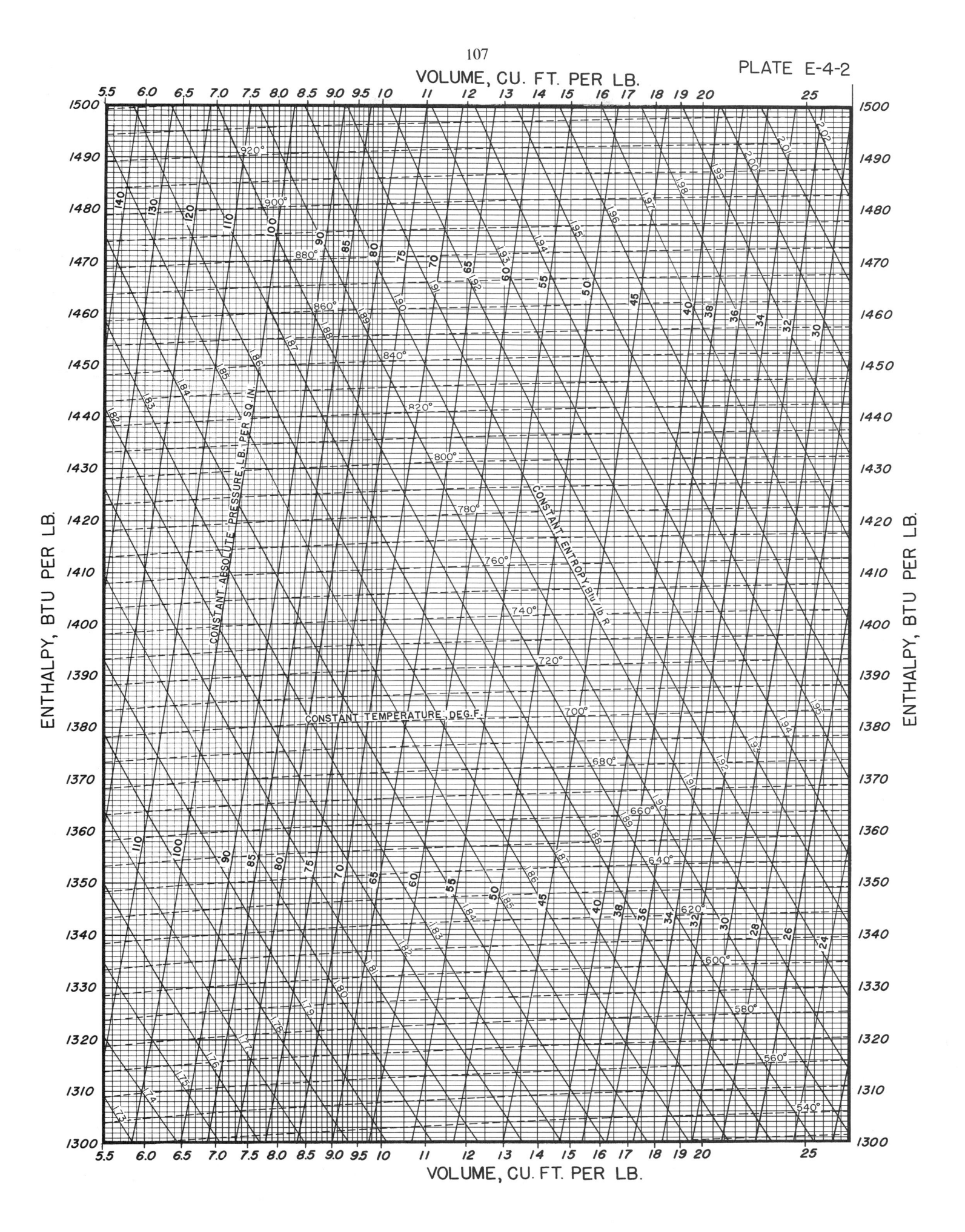

PLATE E-4-3

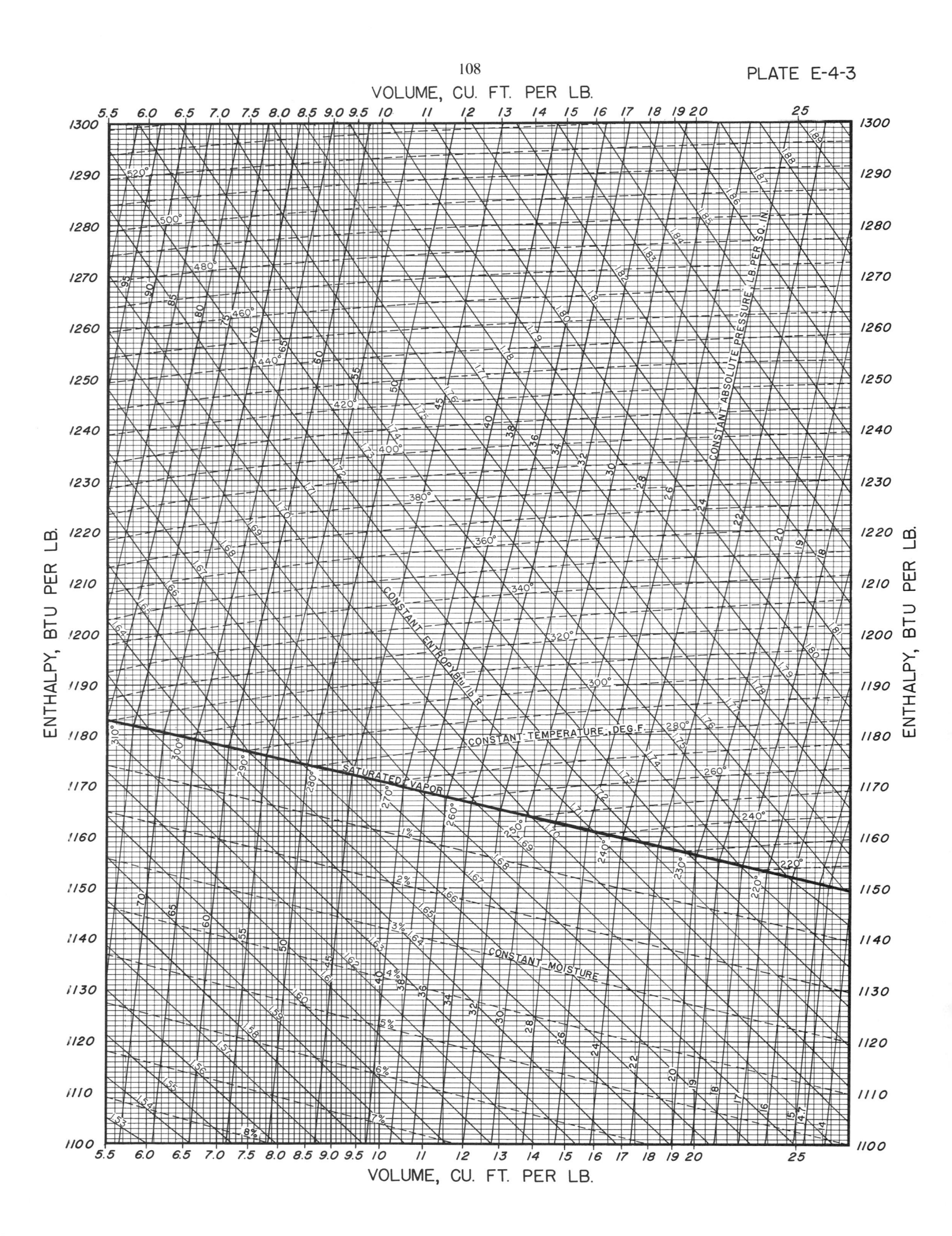

PLATE E-4-4

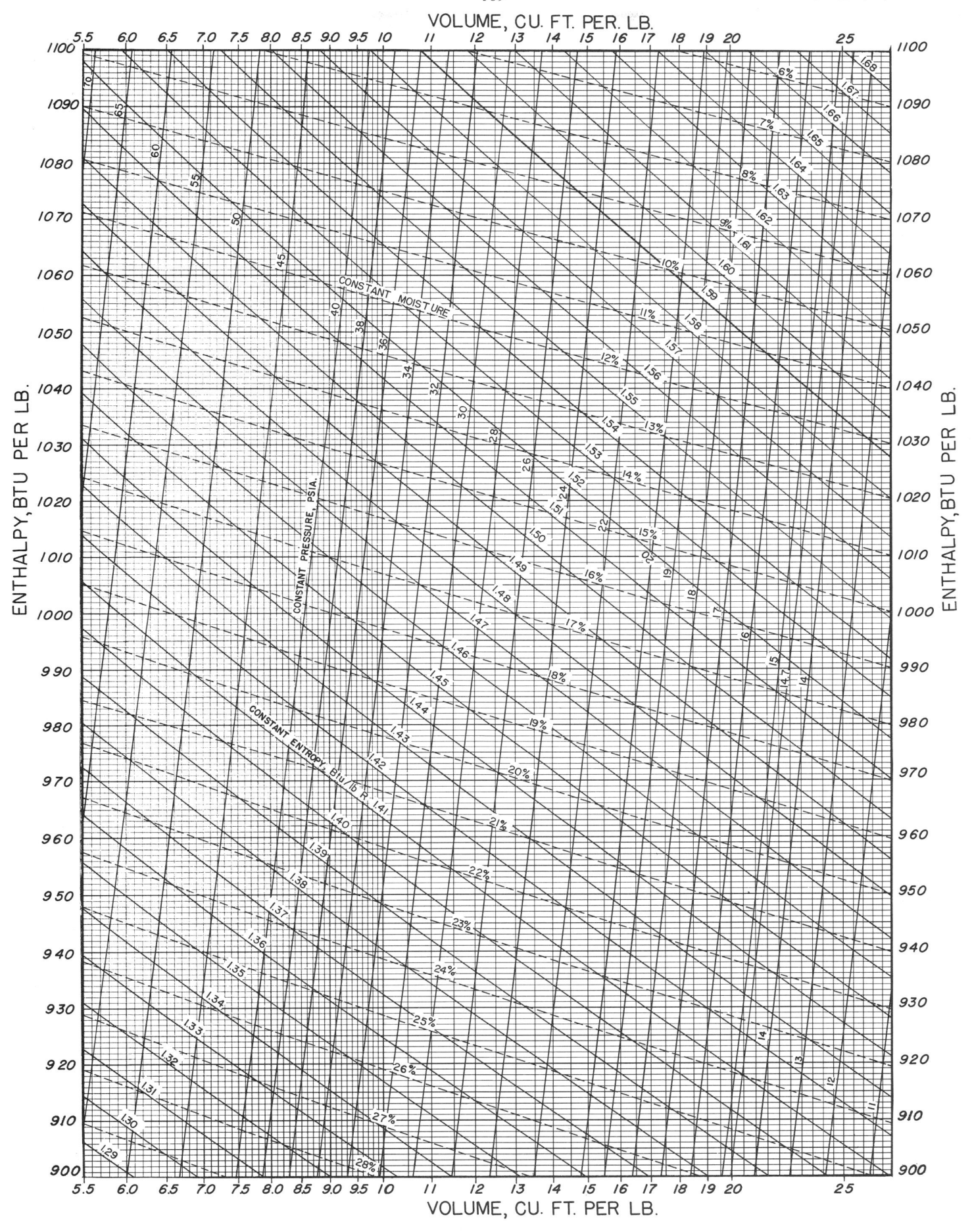

PLATE E-5-1

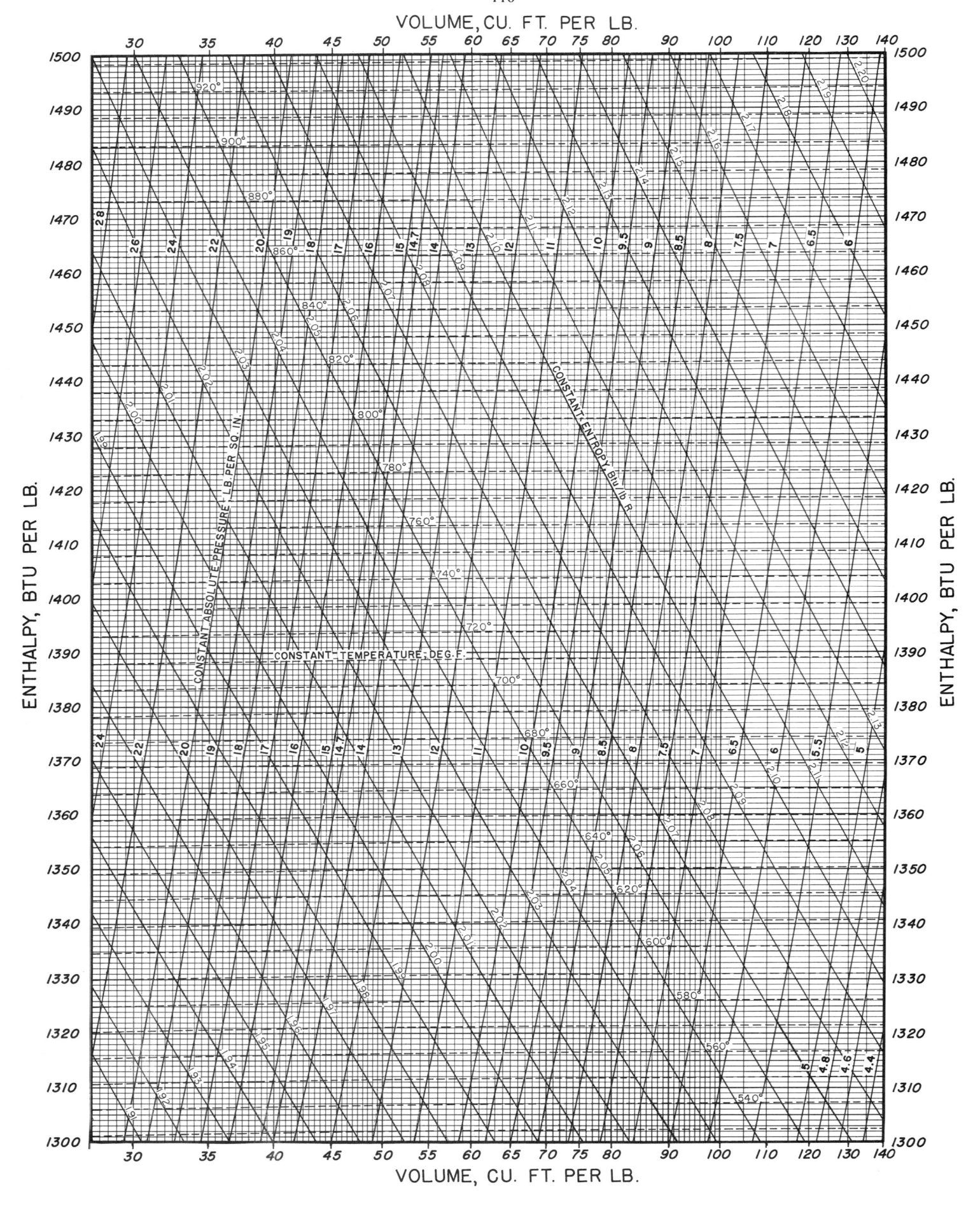

PLATE E-5-2

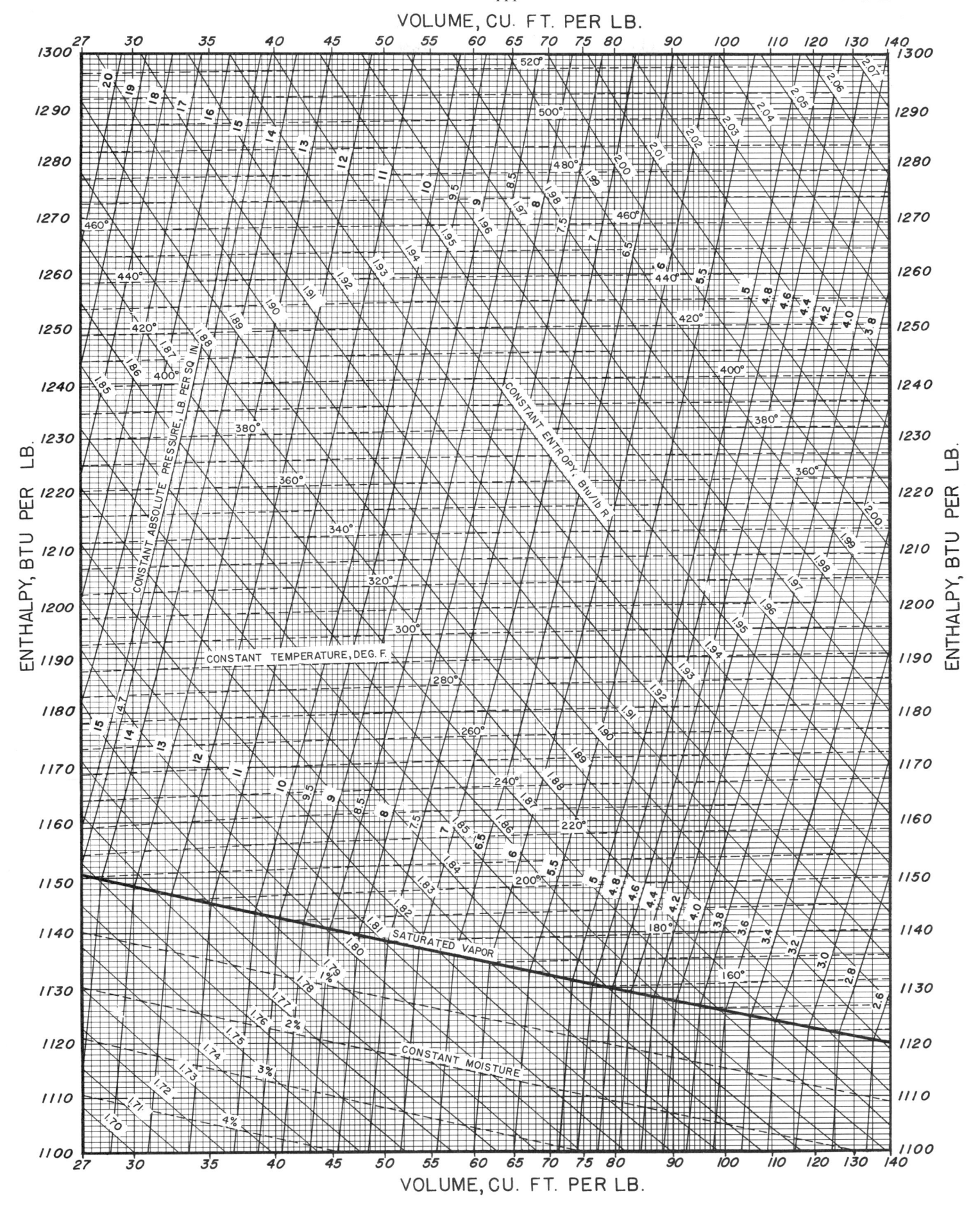

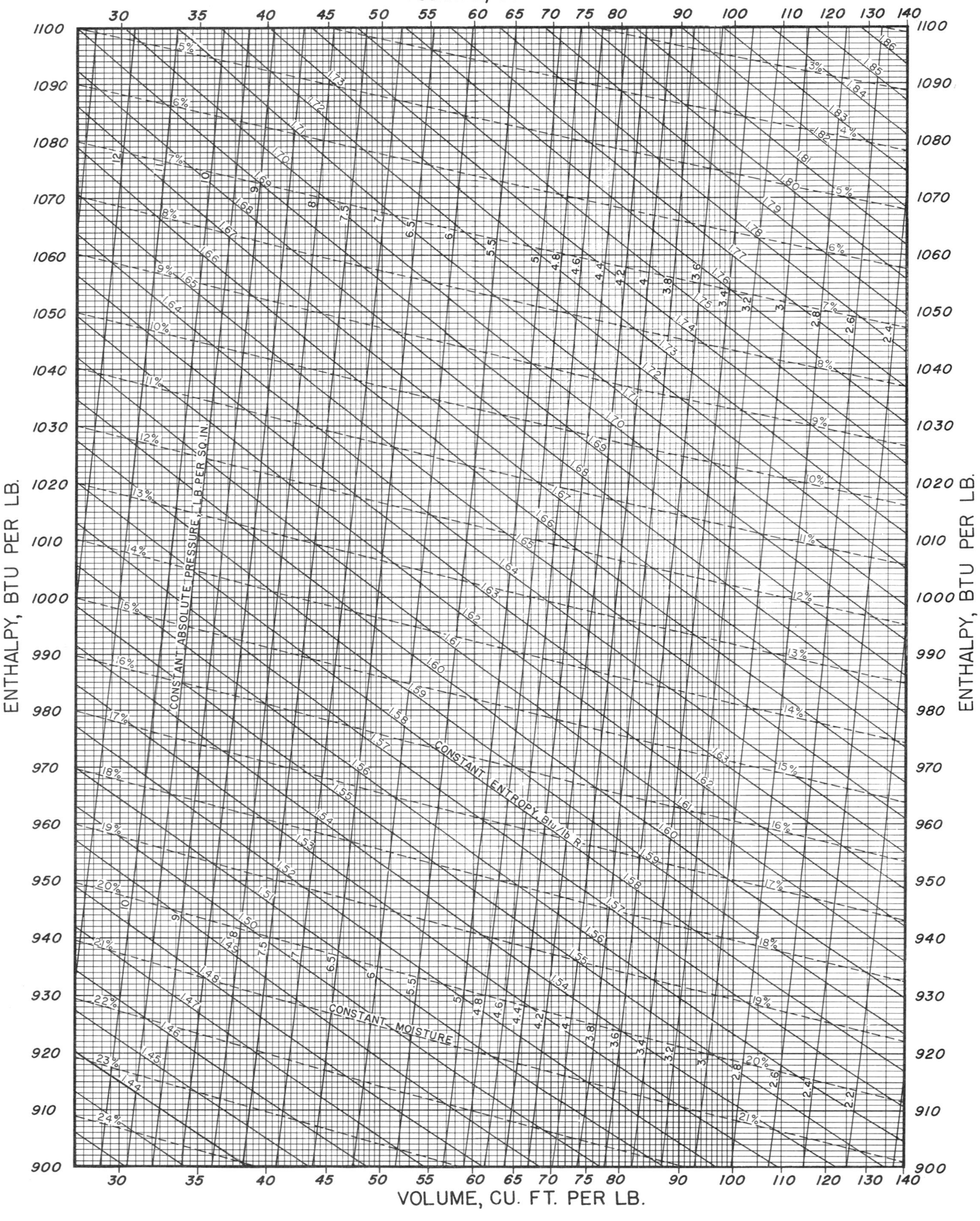
VOLUME, CU. FT. PER LB.
ENTHALPY, BTU PER LB.
CONSTANT ABSOLUTE PRESSURE, LB. PER SQ. IN.
CONSTANT ENTROPY, Btu/lb R
CONSTANT MOISTURE

PLATE E-6-1

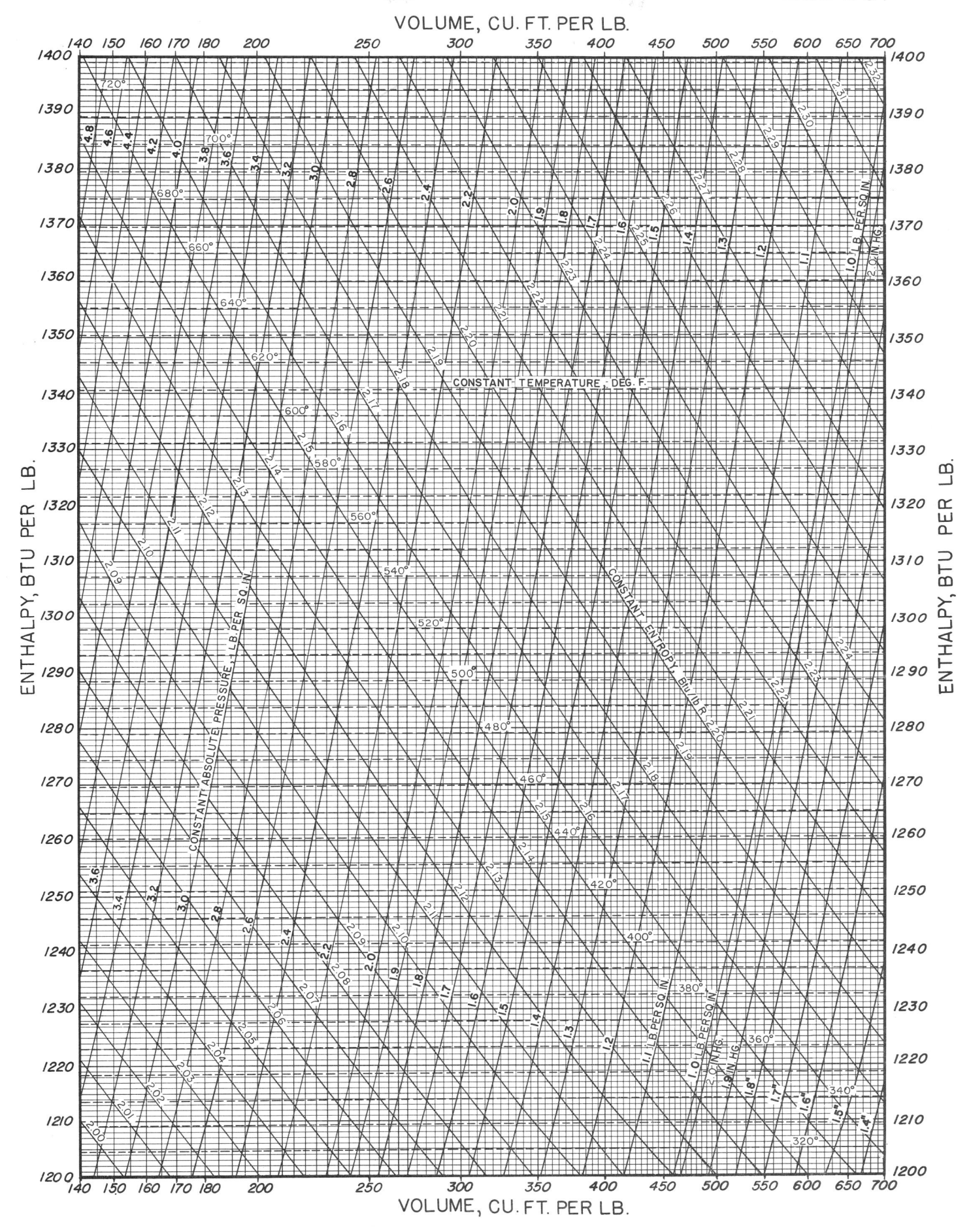

PLATE E-6-2

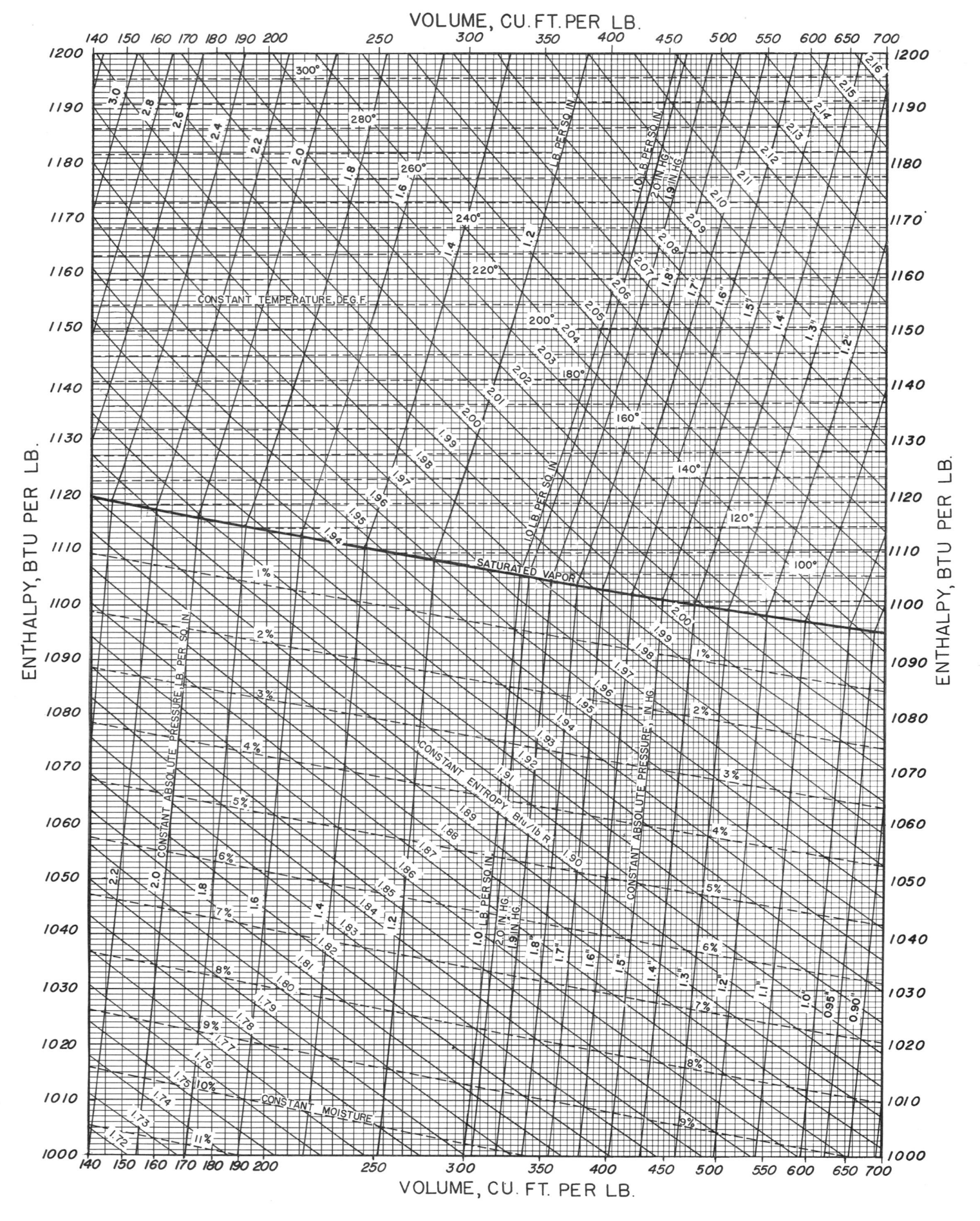

PLATE E-6-3

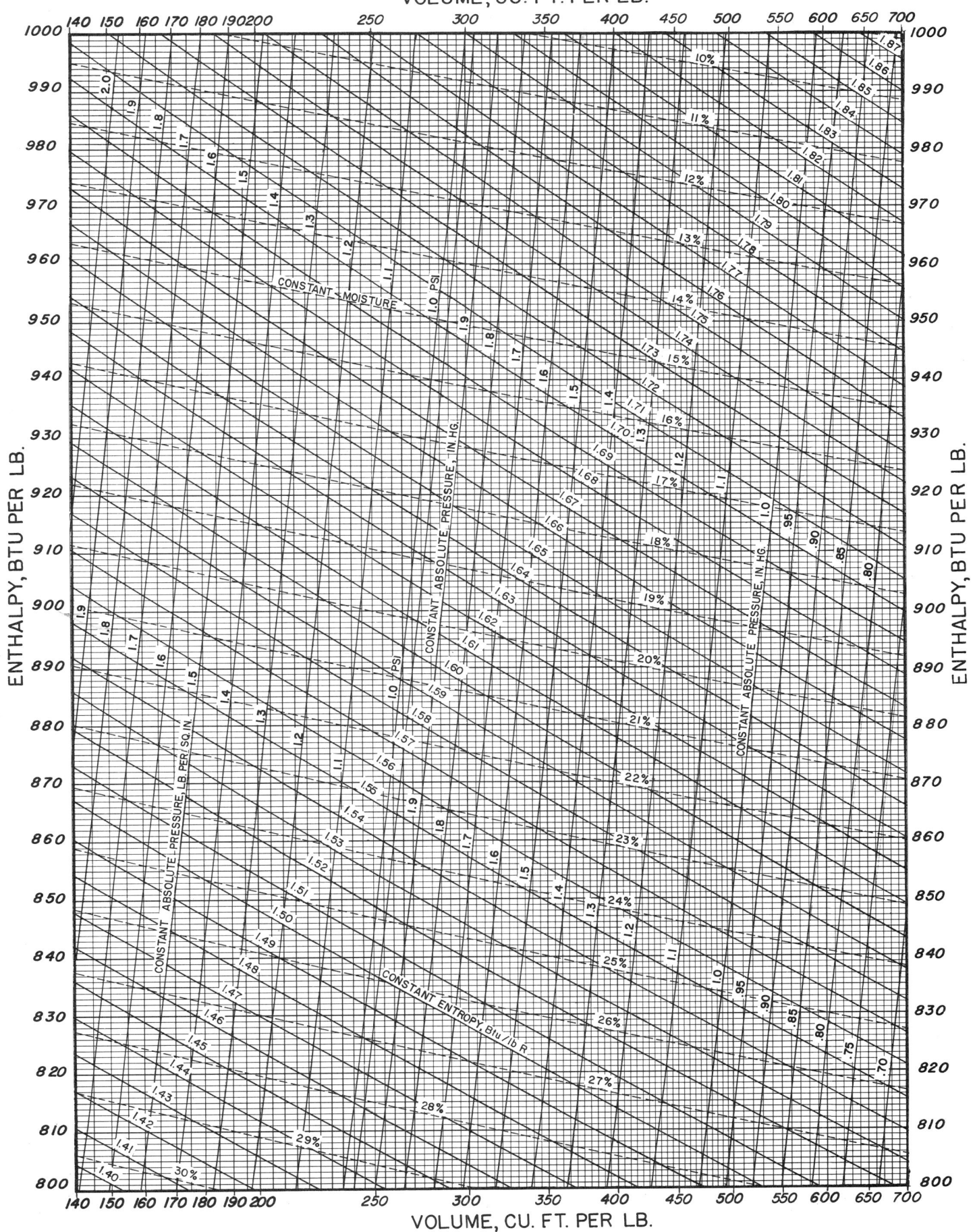

PLATE E-7-1

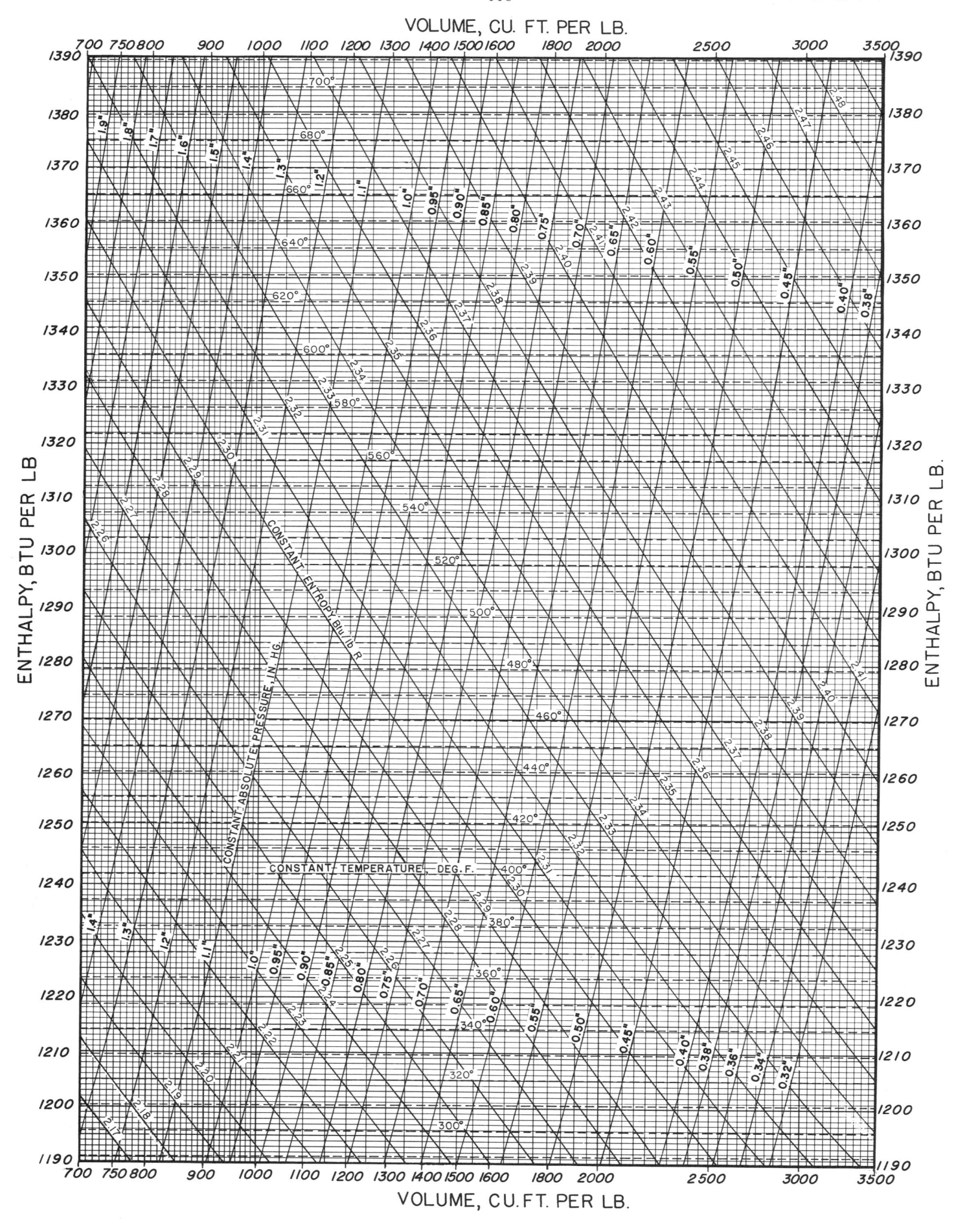

PLATE E-7-2

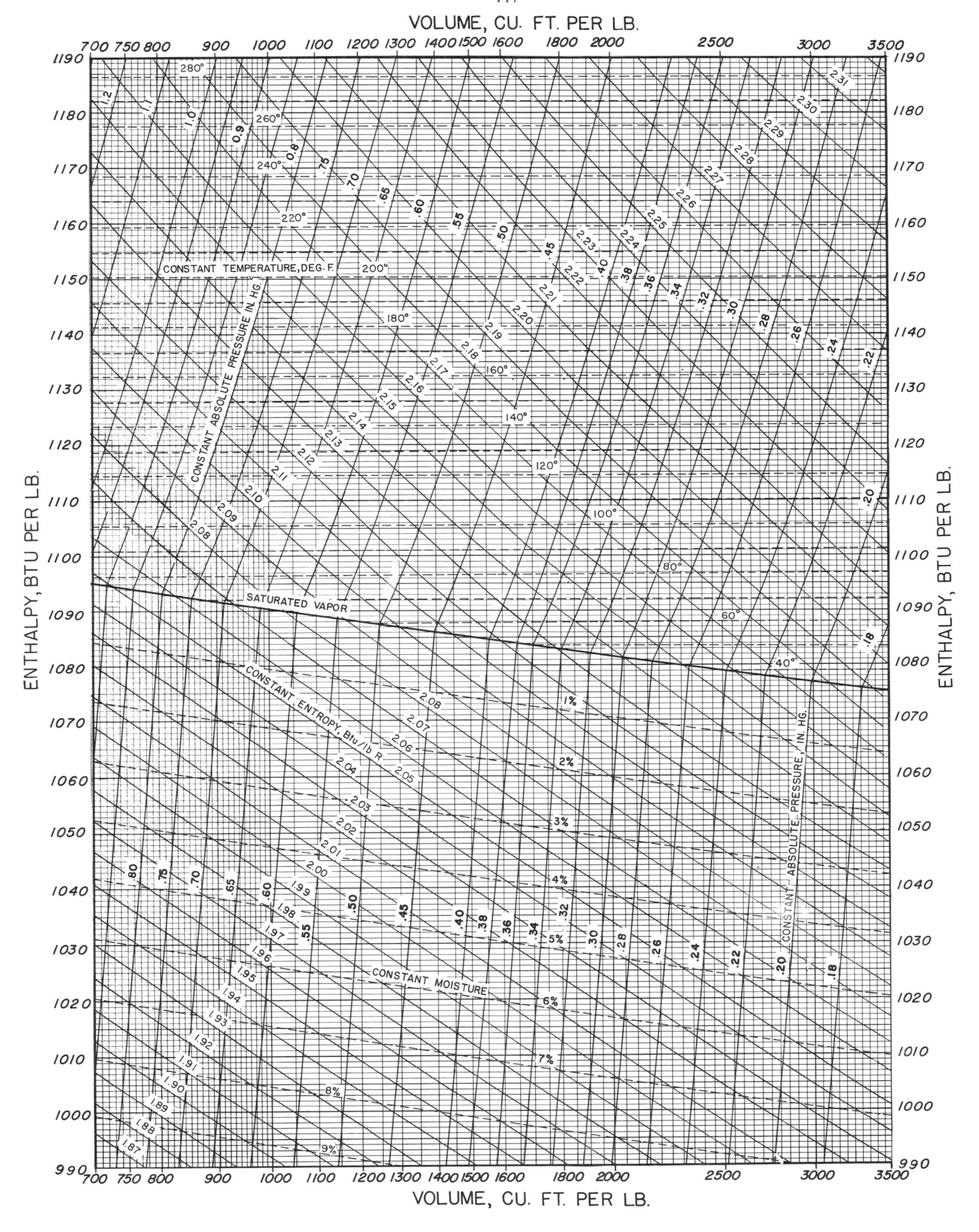

PLATE E-7-3

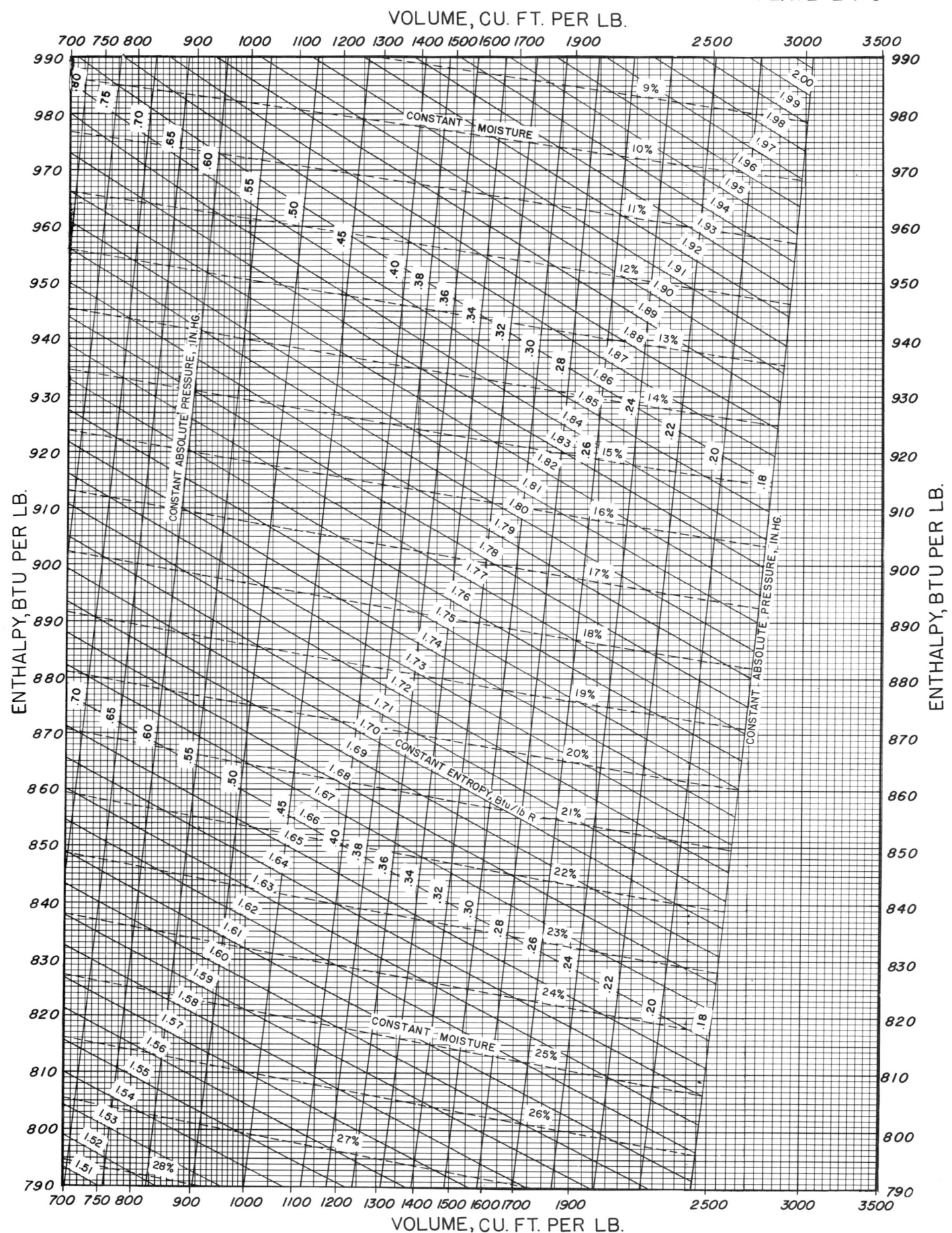

PLATE E-8

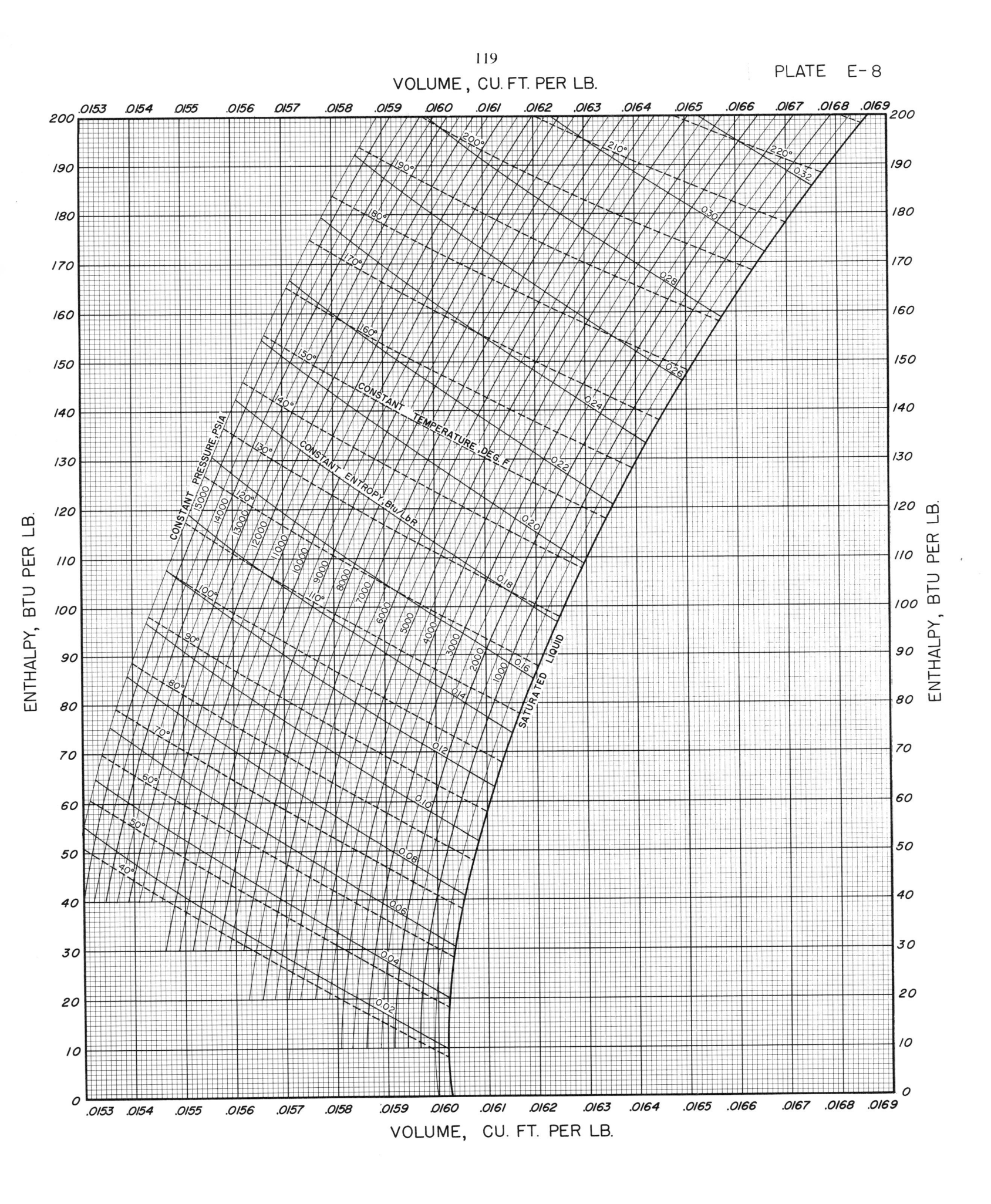

PLATE E-9

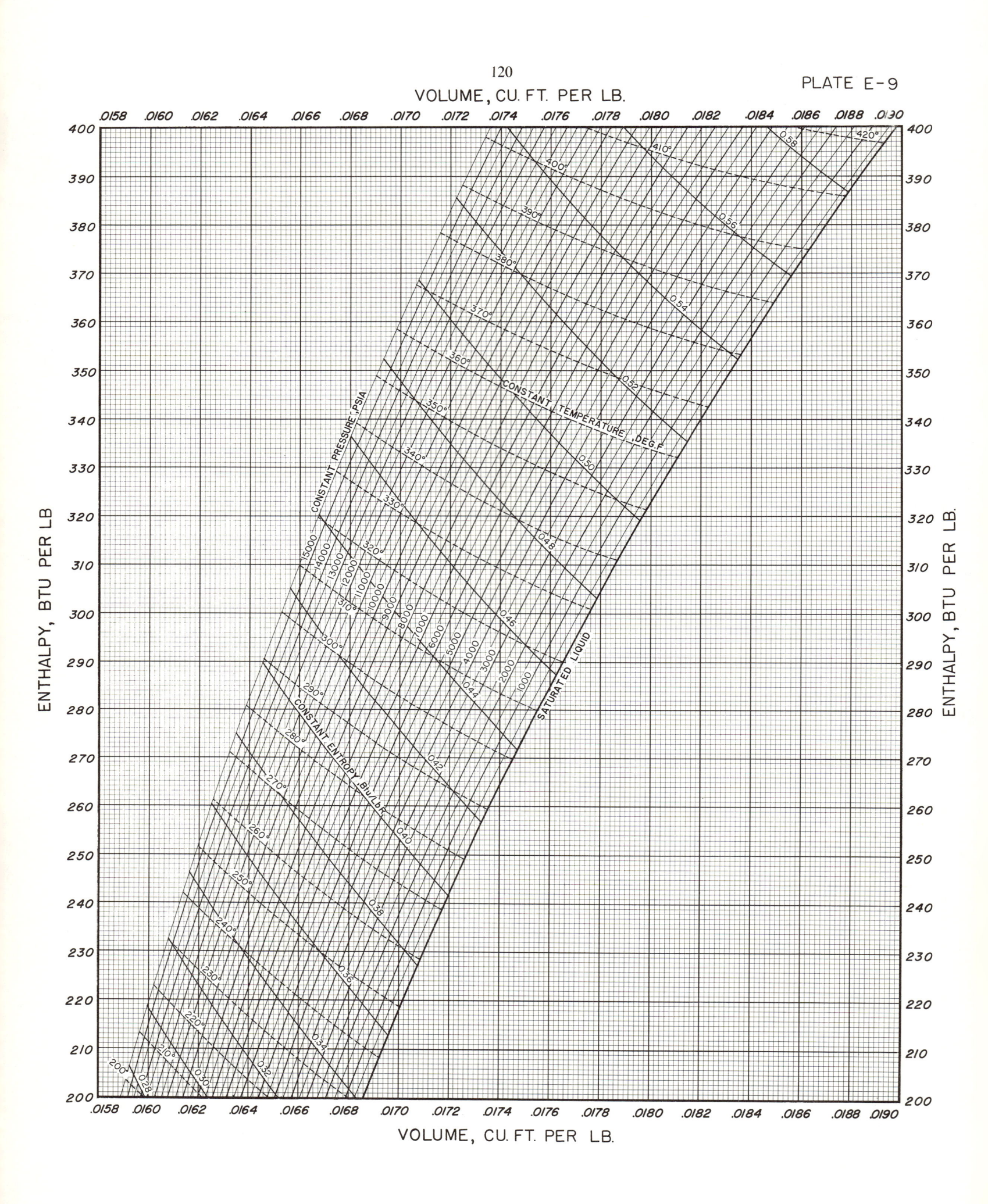

PLATE E-10

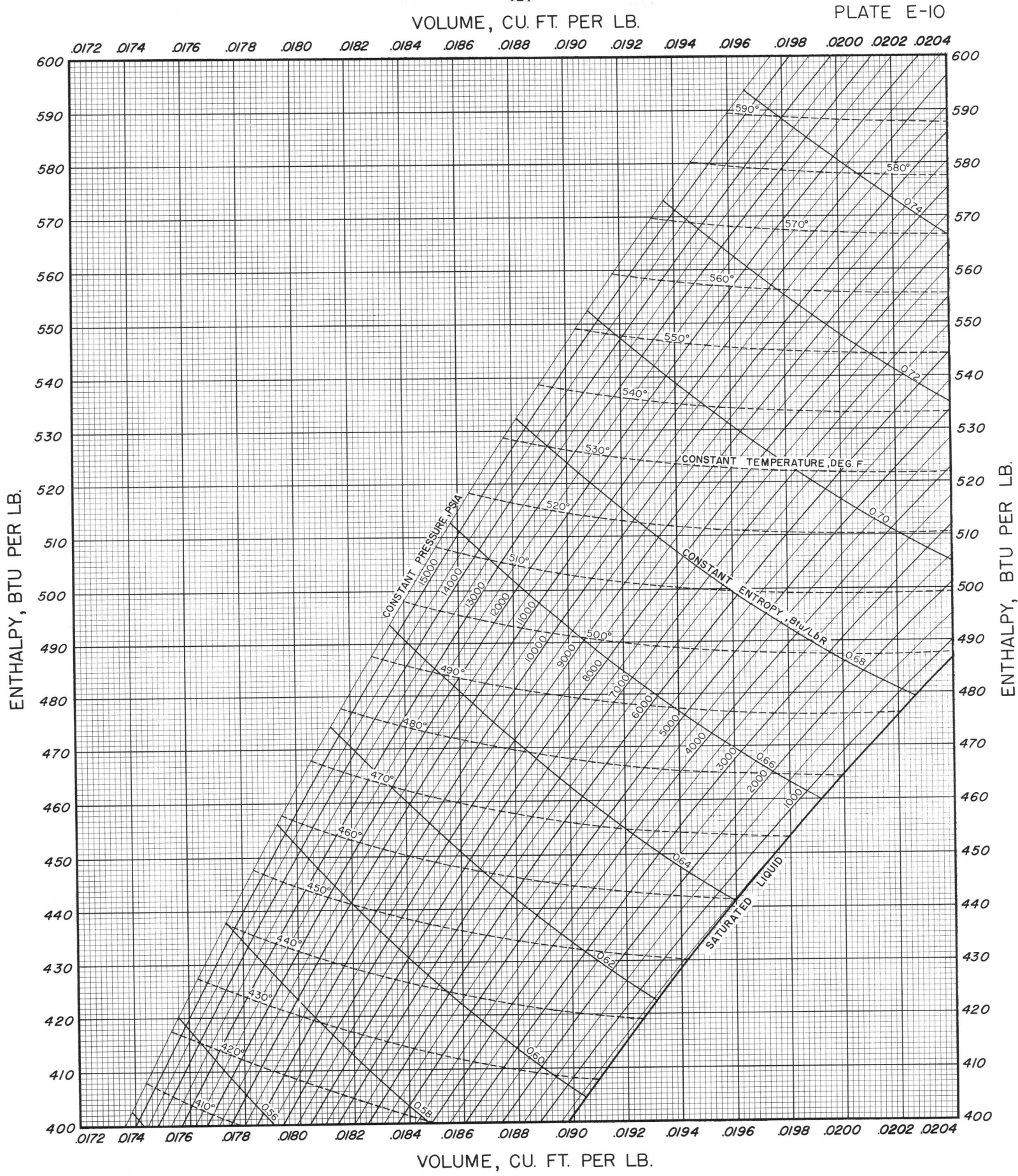

PLATE E-11

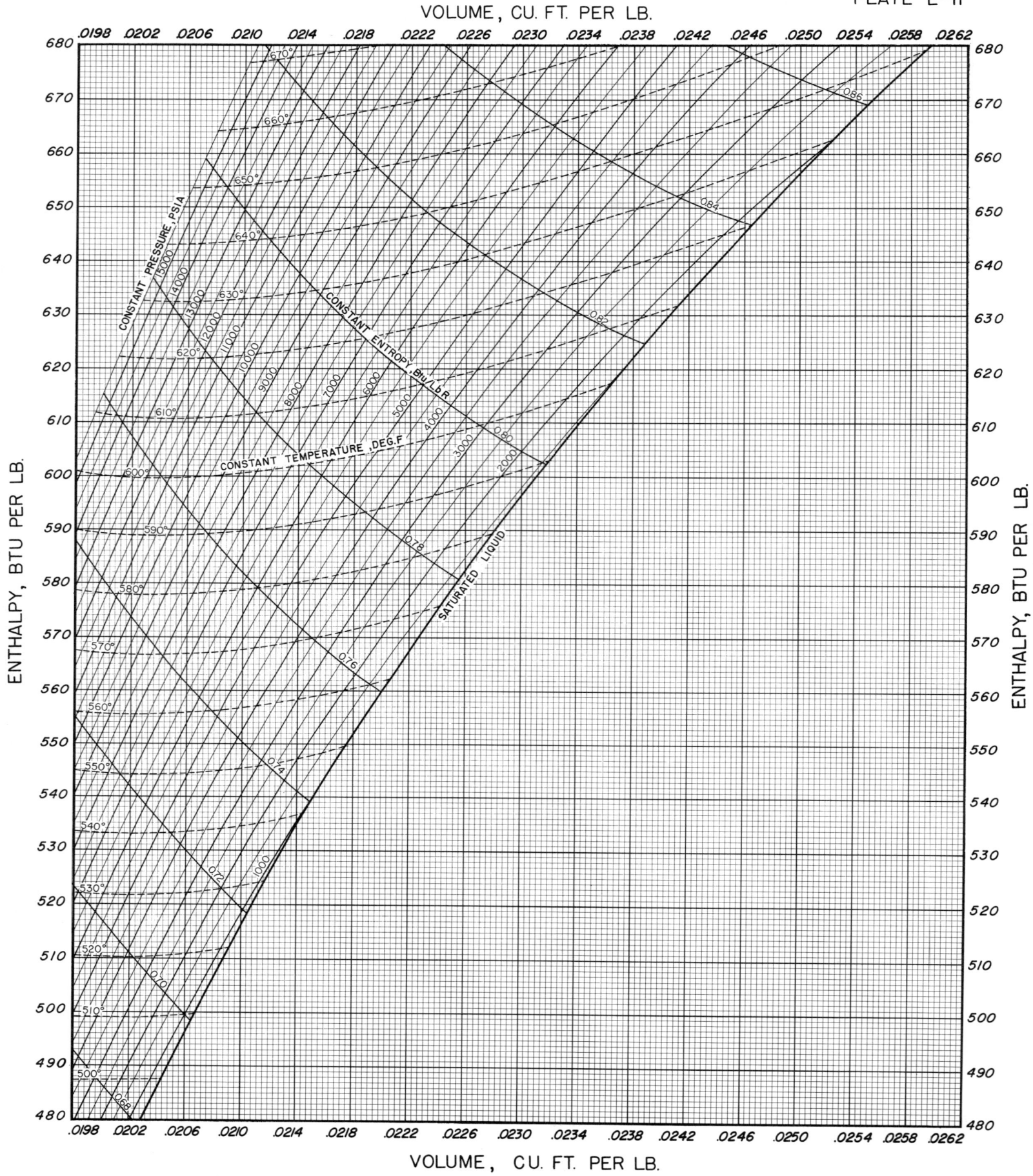

PLATE E-12

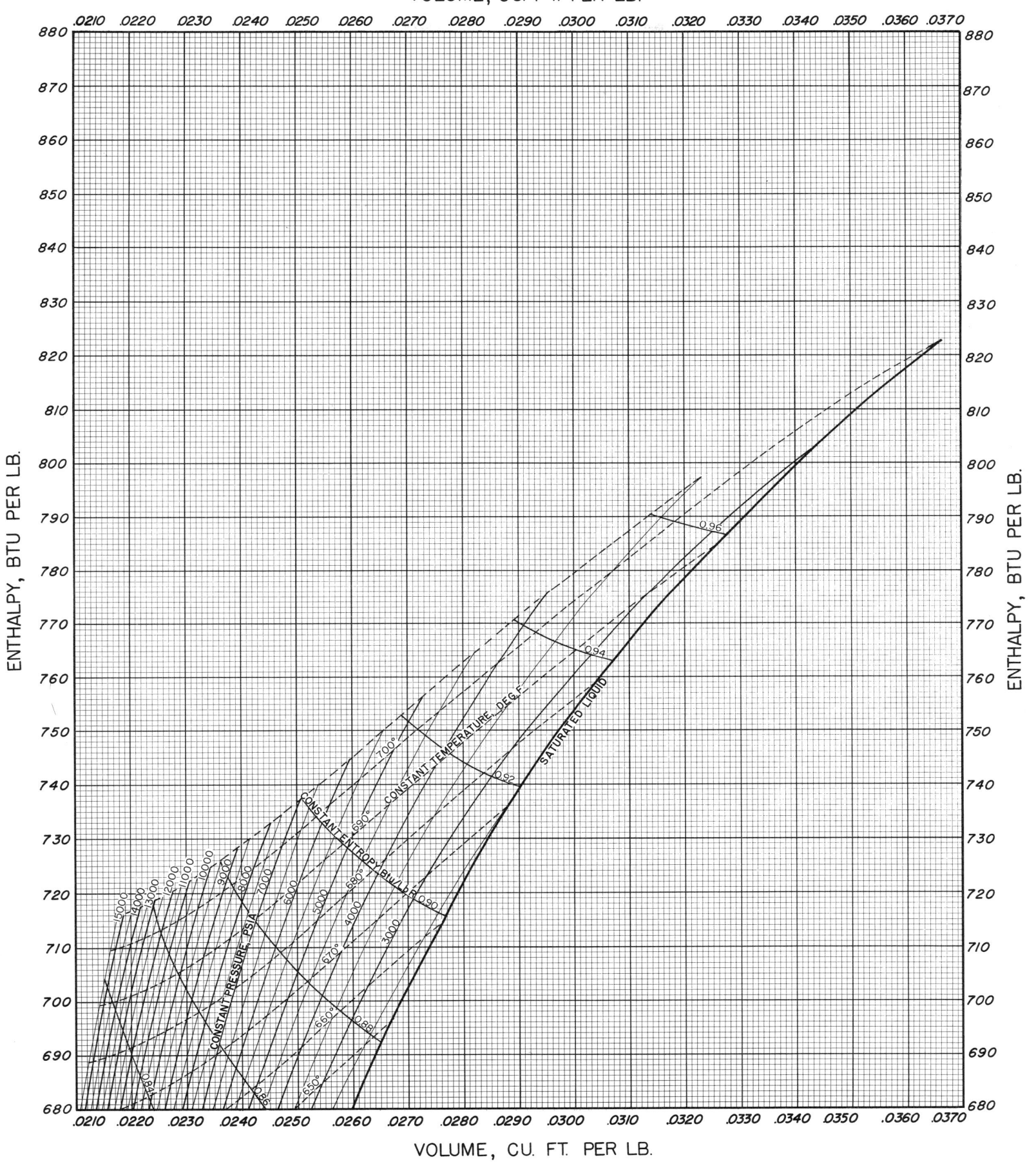

SAMPLE PROBLEMS

Sample Problems

To illustrate the use of the Steam Charts, problems have been solved in both the SI (Metric) and U.S. customary units, covering:

(1) Heat removal from a condenser

(2) Theoretical (isentropic) pump work

(3) Steam calorimeter

(4) Cycle heat rate

PROBLEM (1)

Steam, having a moisture content of 6%, enters a condenser at a pressure of 5.5 kPa (5 500 Pa), and leaves as saturated liquid water. Neglecting kinetic energies, determine the energy removed during condensation.

SOLUTION, SI (METRIC):

From plate M-VI-9 on page 71:
$\quad\quad$ h = 2419.5 kJ/kg $\quad\quad\quad$ t = 35°C

From plate M-VIII on page 87:
$\quad\quad$ h_f = 146.0 kJ/kg

Energy removed = 2419.5 − 146.0 = <u>2273.5 kJ/kg</u>

SOLUTION, U.S. CUSTOMARY:

5 500 Pa × 2.953 × 10^{-4} = 1.625 in Hg a (approximately 1.6 in Hg a)

From plate E-6-2 on page 114:
$\quad\quad$ h = 1040 Btu/lb $\quad\quad\quad$ t = 97°F

From plate E-8 on page 119:
$\quad\quad$ h_f = 65 Btu/lb

Energy removed = 1040 − 65 = 975 Btu/lb

This checks the solution in the SI (Metric) units, allowing for the approximation to the pressure:

975 × 1.055 / (0.454) = <u>2268 kJ/kg</u>

PROBLEM (2)

Under steady flow conditions, 1 kilogram of water is compressed from saturated liquid at 160°C, to a final pressure of 60 MPa (60 × 10⁶ Pa) along an isentropic path. Find the work done, neglecting kinetic energies.

SOLUTION, SI (METRIC):

    From plate M-IX on page 88:
$$h_1 = 676 \text{ kJ/kg} \qquad h_2 = 740 \text{ kJ/kg}$$

    Work is done on the fluid, and is therefore negative

$$W = h_1 - h_2 = 676 - 740 = \underline{-64 \text{ kJ}}$$

SOLUTION, U.S. CUSTOMARY:

    60,000,000 Pa is equivalent to 8700 psia, and 160°C is equivalent to 320°F.

    From plate E-9 on page 120:
$$h_1 = 290 \text{ Btu/lb} \qquad h_2 = 317.5 \text{ Btu/lb}$$

$$\text{Work} = 290 - 317.5 = \underline{-27.5 \text{ Btu/lb}}$$

Converting: (−27.5)(1.055)/(0.454) = −64 kJ

PROBLEM (3)

In a throttling calorimeter the chamber temperature is found to be 120°C at one atmosphere. If the absolute line pressure is 400 kPa (4.0 x 10⁵ Pa), what is the initial quality?

SOLUTION, SI (METRIC):

One atmosphere is equivalent to 101325 Pa

From plate M-V-8 on page 58:
$h_1 = 2716.3$ kJ/kg and $v_1 = 1.77$ m³/kg

From plate M-IV-8 on page 46:
$h_2 = 2716.3$ kJ/kg and $v_2 = 0.453$ m³/kg
at the initial pressure of 4.0×10^5 Pa, the moisture is 1% or 99% quality
This is indicated graphically below:

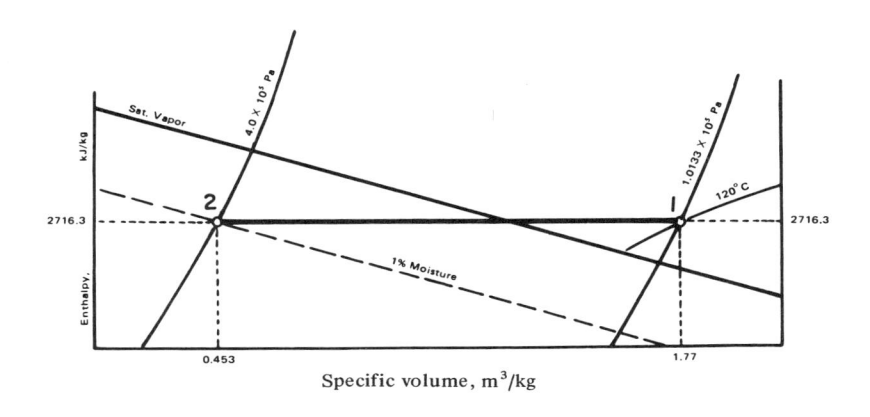

SOLUTION, U.S. CUSTOMARY:

1 atmosphere = 14.696 psia $\qquad t_1 = 120 \times \dfrac{9}{5} + 32 = 248°F$

400,000 Pa is equivalent to 58 psia

From plate E-5-2 on page 111 $\quad h_1 = 1168$ Btu/lb $\qquad v_1 = 28.4$ ft³/lb

From plate E-4-3 on page 108 at 1168 Btu/lb, moisture = 1% or 99% quality

This is indicated graphically below:

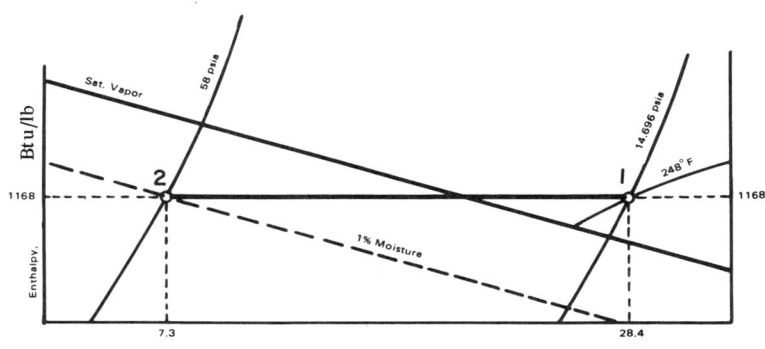

PROBLEM (4)

In a fossil-fueled steam power station, a unit produces 491 000 kW at the generator terminals. Throttle conditions are: Flow = 398 kg/s, pressure 16.65 MPa (16 650 000 Pa), temperature 538°C. Steam enters the reheater with an enthalpy of 3040 kJ/kg, and leaves the reheater at 3.24 MPa, 538°C, with a flow of 354 kg/s. Feedwater enters the economizer at 245°C, 17 MPa (17 000 000 Pa). If the condenser pressure is 5.079 kPa (5079 Pa), find the heat rate of the unit.

SOLUTION, SI (METRIC):

From plate M-II-4 on page 18: @ 16 650 000 Pa, 538°C: h = 3401 kJ/kg

From plate M-III-4 on page 30 @ 3 240 000 Pa, 538°C: h = 3541 kJ/kg

From plate M-X on page 89 @ 17 000 000 Pa, 245°C: h = 1062 kJ/kg
h into reheater (as given in problem)   h = 3040 kJ/kg

(Δh) feedwater to throttle = 3401 - 1062 = 2339 kJ/kg

(Δh) in reheater = 3541 - 3040 = 501 kJ/kg

Turbine heat rate = (398×2339 + 354×501)/491 000 = 2.257 kJ/kW·s
Thermal efficiency = 1/(2.257) = 44.3%

SOLUTION, U.S. CUSTOMARY:

16,650,000 Pa equivalent to 2415 psia   538°C equivalent to 1000°F

From plate E-2-2 on page 99: h = 1461 Btu/lb

17,000,000 Pa equivalent to 2470 psia   245°C equivalent to 473°F

From plate E-10   h = 457 Btu/lb
  (Δh) feedwater to throttle = 1461 - 457 = 1004 Btu/lb
enthalpy entering reheater at 3040 kJ/kg, equiv. to 1307 Btu/lb
reheater output: 3,240,000 Pa equivalent to 470 psia; t = 1000°F
From plate E-3-2 on page 103, h = 1521 Btu/lb
(Δh) reheater = 1521-1307 = 214 Btu/lb
throttle flow = 398 × 3600 / (0.454) = 3,163,000 lb per hr
reheater flow = 354 × 3600 / (0.454) = 2,810,000 lb per hr
Turbine heat rate = (3,163,000 × 1004 + 2,810,000 × 214)/(491,000)
          = 7694 Btu/kW (hr)
Thermal efficiency = 3412/7694 = 44.4%